PUBLISHERS' NOTE

As its title implies, the Series in which this volume appears has two purposes. One is to encourage the publication of monographs on advanced or specialised topics in, or related to, the theory and applications of probability and statistics; such works may sometimes be more suited to the present form of publication because the topic may not have reached the stage where a comprehensive treatment is desirable. The second purpose is to make available to a wider public concise courses in the field of probability and statistics, which are sometimes based on unpublished lectures.

The Series was edited from its inception in 1957 by Professor Maurice G. Kendall, under whose editorship the first 21 volumes in the Series appeared. He was succeeded as editor in 1965 by Professor Alan Stuart.

The publishers will be interested in approaches from any authors who have work of importance suitable for the Series.

CHARLES GRIFFIN & CO. LTD.

GRIFFIN'S STATISTICAL MONOGRAPHS AND COURSES

No. 5, formerly *Characteristic Functions* by E. Lukacs, is now published independently of the Series

For a list of other statistical and mathematical books see back cover.

FAMILIES OF FREQUENCY DISTRIBUTIONS

J. K. ORD

B.Sc. (Econ.), (London), Ph.D. (London), F.S.S.
Lecturer in Statistics, University of Bristol

BEING NUMBER THIRTY OF
GRIFFIN'S STATISTICAL
MONOGRAPHS & COURSES
EDITED BY
ALAN STUART, D.Sc.(Econ.)

1972
HAFNER PUBLISHING COMPANY
NEW YORK

CHARLES GRIFFIN & COMPANY LIMITED
42 DRURY LANE, LONDON, WC2B 5RX

Copyright © J. K. ORD, 1972
All rights reserved

First published 1972

Demy Octavo, viii + 231 pages
13 line illustrations, 23 tables

Set by E W C Wilkins & Associates Ltd London N12 OEH
Printed in Great Britain by Latimer Trend & Co Ltd Whitstable

PREFACE

The study of systems of probability distributions started in 1895 with Karl Pearson's celebrated family of curves, which still plays a major role in modern statistics. Many systems of distributions, both continuous and discrete, have been developed in succeeding years, with a wide range of aims and usage. Applications occur in such diverse fields as economics and linguistics, medicine and sport, to name but a few. Trying to keep track of such a wide-ranging and rapidly expanding literature is probably a hopeless task, but for items of direct interest to statisticians, the reference works of Patil and Joshi (1968) and Johnson and Kotz (1969, 1970) are invaluable (see the bibliography).

My purpose in writing this monograph was not, therefore, to provide an encyclopaedic work, although I hope it will be of some value in this context. Rather, I have tried to look at all the major systems proposed in the literature, to review their main properties, and to discuss the problems involved in selecting and fitting an appropriate model.

The first three chapters cover continuous distributions; Professor Johnson's updating of Elderton's classic work should also be consulted (Elderton and Johnson, 1969). The third chapter on bivariate forms is fairly brief in view of the complementary monograph in this series by Dr Mardia entitled *Families of Bivariate Distributions*, which appeared in 1970. Chapter 4 discusses mixtures of distributions, also providing a link between continuous and discrete systems. Chapters 5–7 consider discrete models, which have grown considerably in importance in recent years, as the conference proceedings edited by Professor Patil (1965, 1970) demonstrate. Finally, in Chapter 8, approximations to various discrete distribution functions are considered and linked with variance-stabilising transforms. There are two appendices; the first provides a brief review of the classical orthogonal polynomials which are used in the main text, while the second lists available tables and charts for various distributions. I am deeply indebted to Alan Mayne for his assistance in compiling this list, and for helpful comments on the manuscript.

I also wish to thank profoundly Professor Alan Stuart for his guidance, first as supervisor, then as editor; without his help and support this project would not have been completed.

My thanks go to all the authors concerned and to the following for permission to reproduce material in the monograph: the *Biometrika*

Trustees, Figures 1.1, 2.1, 2.2, 5.2, 5.3 and 5.4, Tables 2.1, 2.2, 5.1, 5.3 and part of 6.6; Cambridge University Press, Figure 2.3, Table 1.1 and part of Example 1.5; the Editors of *Biometrics* (Table 6.5), *Ecology* (Table 6.6), and the *Journal of the American Statistical Association* (Table 7.3); the Royal Statistical Society, Figures 5.4 and 7.1, and part of Example 5.4; and Professor G.P. Patil, editor of *Classical and Contagious Discrete Distributions*, for Table 6.5.

Finally, my thanks go to Mrs Anne Kempson who patiently and efficiently typed the difficult manuscript and to my wife, Rosemary, for her help in compiling the references and looking after one family while I dealt with the others.

J.K.O.

BRISTOL
February, 1971

CONTENTS

CHAPTER 1

THE PEARSON SYSTEM AND THE EXPONENTIAL FAMILY OF CURVES

1.1 Introduction

Towards the end of the nineteenth century, when interest in fitting curves to data was rapidly growing, it became apparent that samples from many sources could show distinctly non-normal characteristics, and that this was not due (entirely!) to faulty measurement, but was an inherent feature of the population.

Karl Pearson and his associates at University College London were to the fore in these studies (see E.S. Pearson 1965, 1967, for a historical account of events at this time), and the search for acceptable alternatives to the assumption of normality led to the development of the Pearson system of curves. However, not all were convinced of the need for curves other than the normal, as shown in Pearson's reply (1905) to Ranke and Greiner, who felt that asymmetric curves were largely unnecessary.

Nevertheless, by the turn of the century, most informed opinion had accepted that populations might be non-normal, and various schemes were being proposed to generate systems of curves which, if they included the normal, did so only as a special case. The scope of such systems varies and their usefulness can only be decided in the light of their properties and subsequent usage.

Most early writers employed monotone transformations of the random variable X, assuming the new variable $Y = g(X)$ to be normally distributed, that is, with density function, when $Y = y$,

$$\alpha(y) \propto \exp[-\tfrac{1}{2}\{(y - \mu)/\sigma\}^2] \qquad (1.1)$$

(which we adopt as a standard notation); we also speak of Y being $N(\mu, \sigma)$.

Early examples are:

(i) Galton (1879), McAlister (1879): $y = \ln(x + a)$

(ii) Kapteyn (1903): $y = (x + a)^q$

(iii) Edgeworth (1898):   general form of $g(x)$

(iv) Fechner (1897): $y = (x - \mu)/\sigma_1, \quad x \geqslant \mu$

$\qquad\qquad\qquad = (x - \mu)/\sigma_2, \quad x < \mu,$

the double-Gaussian curve.

A modern approach to transformation, or translation, methods is given by Johnson (1949a,b) whose work is discussed in sections 2.12 *et seq.*

Karl Pearson's (1895) approach was essentially different. Observing that the normal density function obeyed the differential equation

$$\frac{d\alpha}{dx} = \frac{(x - a)\,\alpha(x)}{b}, \tag{1.2}$$

where $a = \mu$, $b = -\sigma^2$, and that the hypergeometric distribution for direct sampling without replacement obeyed a difference equation of the form

$$\Delta f_{r-1} = f_r - f_{r-1} = \frac{(r - a)\,f_r}{b_0 + b_1 r + b_2 r^2}, \quad r \geqslant 1 \tag{1.3}$$

where f_r is the probability of r successes and a, b_i are parameters, he used these forms to generate his system. Allowing the lattice width on which r is defined to diminish, we arrive, in the limit, at the differential equation, for density function $f(x)$,

$$\frac{df}{dx} = \frac{(x - a)\,f(x)}{b_0 + b_1 x + b_2 x^2}. \tag{1.4}$$

Whatever current views on the value of curve fitting may be, no-one can deny that this equation had a considerable impact on the course of statistical theory; although, with the benefit of hindsight, one may feel that the subject did not move entirely in the right direction. We discuss the Pearson system in section 1.3 and the following sections.

A third approach is to assume that an arbitrary density function, $h(x)$ say, can be represented by a series based on the normal density function, that is

$$h(x) = \sum_{j=0}^{\infty} c_j H_j(x)\,\alpha(x), \tag{1.5}$$

where the $H_j(x)$ are polynomials of order j in x. The form of the series depends on the coefficients c_j which depend, in turn, on the genesis of (1.5). Our discussion of series expansions starts in section 2.1.

The central place occupied by the normal distribution in statistical theory has meant that most systems of distributions have the normal curve as their core. While this is useful when approximating to sampling distributions, it is not a necessary constraint. For example, Romanovsky (1924) used orthogonal polynomials based on members of the Pearson system (see Appendix A); while Burr (1942) employed a differential equation involving distribution functions (see section 2.16).

1.2 The need for systems of distributions

The importance of finding a suitable curve to fit data has some-what declined. A modern researcher proceeds by building a model of the process he is studying and then testing his data against the model. The advent of modern high-speed computers has relaxed the constraint of computational feasibility (subject to size of research grant), enabling more complex models to be tested.

Nevertheless, these systems continue to be useful. Their value to the modern statistician is summarized by E.S. Pearson (1961):

> "(1) in representing the sampling distribution of statistics for which only the moments or cumulants are known or readily expressible;
>
> (2) as providing a series of typical forms of non-normal parent distributions, which may be used to explore the 'robustness' of standard tests, either mathematically or by Monte Carlo methods."

The various systems also provide suitable classes for theoretical studies, the Pearson system being important in this respect as it includes the t, χ^2 and F distributions.

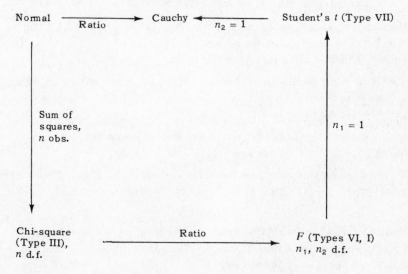

1.3 The Pearson system of curves

In addition to the original papers of Karl Pearson (1895, 1901, 1916) to be found in the collection edited by E.S. Pearson (1948), the basic theory of the differential equation system and its properties are discussed in many works, notably Elderton and Johnson (1969), and Kendall and Stuart (1969). We therefore content ourselves with a review of the main properties and a summary of results in Table 1.1 (pages 6–7).

To simplify the discussion without losing any major results, we assume the density function to be at least once differentiable in the range of interest of the random variable $[u, v)$ say, where u and/or v may be infinite.

The denominator of the differential equation (1.4) has two roots which may be

(a) real with the same sign,

or (b) real with opposite signs,

or (c) both imaginary.

If we write $\kappa = b_1^2/4b_0 b_2$, then the above three cases are Pearson's *Main Types*, which, following Elderton's notation (see Elderton and Johnson, 1969), are (a) Type VI, $\kappa > 1$; (b) Type I, $\kappa < 0$; (c) Type IV, $0 < \kappa < 1$. These curves have one, two and no finite end-points respectively. The main types follow when the parameters in (1.4) are unrestricted, except insofar as is necessary to ensure that $f(x)$ is a density function with total measure 1. By inserting particular values of the parameters into (1.4) we may establish

Property P1. If a curve is contained in the system, so is the limiting form which is obtained as the parameters approach a particular set of values. These limiting forms are the *transitional types*.

Throughout the list of properties, proofs are given only where the results are not obvious.

Property P2. From (1.4), $df/dx = 0$ has at most one root, when $x = a$; this is a maximum (mode) or minimum (antimode) according as $(df/dx)_{x=a-\delta}$ is positive or negative for small $\delta > 0$.

Just as we may constrain the parameter values to obtain the transitional types, we may restrict the range of the random variables, so that

Property P3. If a distribution defined for $x \in [u, v)$ is contained in the system, any truncated form of that distribution, defined for $x \in [u^*, v^*)$, $u \leqslant u^* < v^* \leqslant v$, is in the system.

To relate a particular "type" to a set of parameter values, it is necessary to express the parameters of the differential equation in terms of the parameters of that type; to do this Pearson used the moments of the distributions, leading to his Method of Moments for fitting a curve to data.

From (1.4) we have

$$\int x^j (b_0 + b_1 x + b_2 x^2) \, df = \int x^j (x - a) \, f dx, \tag{1.6}$$

the integrals being defined over the entire range of x, provided both sides exist.

Property P4. Provided the $(j + 1)$th non-central moment, μ'_{j+1}, exists, the moments satisfy the recurrence relation

$$\{(j + 2)b_2 - 1\}\mu'_{j+1} + \{(j + 1)b_1 - a\}\mu'_j + jb_0\mu'_{j-1}$$
$$= [x^j (b_0 + b_1 x + b_2 x^2) f]_u^v. \tag{1.7}$$

If there is sufficiently high order contact at both ends of the range, the term on the right-hand side vanishes. Taking the origin at the mean gives a relation for the central moments; for a relation between the cumulants, see Exercise 1.1.

Relation (1.7) enables us to find the central moments, μ_j, and the Pearsonian coefficients $\beta_1 = \mu_3^2/\mu_2^3$, $\beta_2 = \mu_4/\mu_2^2$. When the term on the right-hand side of (1.7) does vanish, the parameters of (1.4) may be expressed, on taking the origin at the mean, as

$$\left. \begin{array}{l} a = b_1 = \beta_1^{1/2} \sigma (\beta_2 + 3)/D \\ b_0 = -\sigma^2 (4\beta_2 - 3\beta_1)/D \\ b_2 = (2\beta_2 - 3\beta_1 - 6)/D \end{array} \right\} \tag{1.8}$$

where $\mu_2 = \sigma^2$, $D = 10\beta_2 - 12\beta_1 - 18$.

Therefore, provided the first four moments exist, we can characterise a particular distribution by these values; while κ, defined above, may be written as

$$\kappa = \frac{\beta_1 (\beta_2 + 3)}{4(2\beta_2 - 3\beta_1 - 6)(4\beta_2 - 3\beta_1)}. \tag{1.9}$$

Pearson used the statistic

$$\lambda = \{(1 - b_2)\kappa - 4b_2\}/4b_2^2 (1 + \kappa) \tag{1.10}$$

to check whether a J-shaped curve was appropriate (his Types VII, IX, XI); he also proposed

$$sk = (\text{mode} - \text{mean})/\sigma \tag{1.11}$$

as a measure of skewness (see Table 1.1).

Table 1.1 *Curves in*

Type	Equation
MAIN	
I	$(1 + x/a_1)^{m_1} (1 - x/a_2)^{m_2}$
IV	$(1 + (x/a)^2) \exp\{-\nu \tan^{-1}(x/a)\}$
VI	$(x - a)^{m_2} x^{-m_1}$
TRANSITION NORMAL	$e^{-\frac{1}{2} x^2}$
II (I)	$(1 - (x/a)^2)^m$
VII (IV)	$(1 + x^2/a^2)^{-m_1}$
III	$(1 + x/a)^m e^{-\gamma x}$
V	$x^{-m} e^{-\gamma/x}$
VIII (I)	$(1 + x/a)^{-m}$
IX (I)	$(1 - x/a)^m$
XII (I)	$\left(\dfrac{(3 + \beta_1)^{\frac{1}{2}} + \beta_2 + x/\sigma}{(3 + \beta_1)^{\frac{1}{2}} + \beta_2 - x/\sigma}\right)^{[\beta_1/(3 + \beta_2)]^{\frac{1}{2}}}$
X (III)	$e^{-\gamma x}$
XI (VI)	x^{-m}

where $C = 2\beta_2 - 3\beta_1 - 6$, $D = 10\beta_2 - 12\beta_1 - 18$.

$\bar{U}$: both tails infinite $\bar{U}.1$: one tail infinite L: both tails finite

the Pearson system

Origin	Criteria	Comments
Mode	$\kappa < 0$	L; skew, B, U, J or J'
$\nu a/(2m-2)$ after mean	$0 < \kappa < 1$	$\bar{U}$; skew, B
a before start	$\kappa > 1$	$\bar{U}.1$; skew, B or J $(m_1 > m_2 + 1)$
Mode (mean)	$\kappa = \beta_1 = 0,\ \beta_2 = 3$	$\bar{U}$; symmetric, B
Mode (mean)	$\kappa = \beta_1 = 0,\ \beta_2 < 3$	L; symmetric, B. (U if $\beta_2 < 1.8$)
Mode (mean)	$\kappa = \beta_1 = 0,\ \beta_2 > 3$	$\bar{U}$; symmetric, B
Mode	$\kappa \to \infty,\ \ C = 0$	$\bar{U}.1$; skew, B or J.
Start of curve	$\kappa = 1$	$\bar{U}.1$; skew, B.
End of curve	$\kappa < 0,\ \lambda = 0,\ D < 0$	L; J
End of curve	$\kappa < 0,\ \lambda = 0,\ D > 0$	L; J
Mean	$\kappa < 0,\ \ \ \lambda = D = 0$	L; J'
Start of curve	$\kappa \to \infty,\ \beta_1 = 4,\ \beta_2 = 9$	$\bar{U}.1$; J
a before start	$\kappa > 1,\ \lambda = 0,\ C > 0$	$\bar{U}.1$; J Pareto distribution

B: bell-shaped J: J-shaped J': twisted J-shaped U: U-shaped.

(Reproduced by permission from Elderton and Johnson, 1969)

1.4 Further properties of the Pearson system

A relation for the characteristic function, defined as

$$\phi(t) = \int e^{\theta x} f(x)\, dx, \quad \theta = it,$$

integration taken over the range of x, may be obtained in the same way as for the moments. We assume that $\phi'(= d\phi/d\theta)$ and ϕ'' exist for some θ in an interval $\theta_L < \theta < \theta_U$.

Property P5 (Kendall, 1941). Assuming high order contact at the ends of the range, the characteristic function of a distribution in the Pearson system satisfies the equation

$$b_2\theta\phi'' + (1 + 2b_2 + b_1\theta)\phi' + (a + b_1 + b_0\theta)\phi = 0. \qquad (1.12)$$

An important feature of Pearson curves relating to their points of inflexion was discovered by Zoch (1935). Provided that $f(x)$ is twice differentiable on the interior of the range of x, the roots of $f''(x) = 0$ are

$$x = a \pm \{a^2 + (ab_1 + b_0)/b_2\}^{\frac{1}{2}} \qquad (1.13)$$

or $\qquad x = \pm (b_0/b_2)^{\frac{1}{2}}$, if the origin is at the mode.

We thus have

Property P6. There may be 0, 1 or 2 points of inflexion, depending on the form of curve and the range of x, but if there are two such points, they must be equidistant from the mode.

The mean deviation of the curves, provided μ_1', exists, is given by

$$M = 2\int_{\mu}^{\infty} |x - \mu|\, f dx. \qquad (1.14)$$

Property P7 (Pearson, 1924c). The mean deviation for the system is

$$M = 2b_0 f(\mu)/(1 - 2b_2) = \frac{12\mu_2 f(\mu)(\beta_2 - \beta_1 - 1)}{(4\beta_2 - 3\beta_1)}. \qquad (1.15)$$

As observed by Johnson (Ramasubban, 1958), the latter reduces to $M = 2\mu_2 f(\mu)$ when $2\beta_2 - 3\beta_1 - 6 = 0$, which from Table 1.1 implies that the distribution is either normal or Type III.

1.5 Extensions of the Pearson system

The Pearson differential equation is of the form

$$\frac{d \ln f}{dx} = N(x)/D(x), \qquad (1.16)$$

where N, D are polynomials in x of order n, d respectively. Pearson discussed $n = 1$, $d = 2$, and it was inevitable that curves corresponding to higher n, d values would be investigated. The curves of these higher-order systems can, at least in principle, be identified by their moments, provided enough exist.

(i) $n = 1$, 2; $d = 3$. Work on these forms by Heron is mentioned in Pearson (1916), though it remained unpublished. This was later taken up by Mouzon (1930) and Zoch (1934). These curves may be bimodal when $n = 2$.

(ii) $n = 1$; $d = 4$. Hansmann (1934) considered the symmetric curves generated by a quartic denominator; these curves allow higher β_1, β_2 values and a variety of U-shaped curves. He employed a β_2, β_4 $(= \mu_6/\mu_2^3)$ diagram to show the region defined by each curve.

(iii) $n = 1$; any d. The general form decomposes into partial fractions, and $f(x)$ consists of three types of term:

$$(x^2 + ax + b), \quad (x + a), \quad \exp\{\tan^{-1}(x + a)/b\}.$$

The curves are of some theoretical interest, but data fitting would be exceedingly difficult and interpretation well-nigh impossible.

(iv) We could allow a single discontinuity, at $x = 0$; that is, we consider

$$\frac{d \ln f}{dx} = \frac{c_0(x - a)}{b_0 + b_1 x + b_2 x^2}, \tag{1.17}$$

where $\quad c_0 = 1 - c + 2c\, S(x),$ \hfill (1.18)

and $\quad S(x) = 1$ for $x \geqslant 0$,

$\qquad\qquad\quad = 0$ for $x < 0$.

This generates a system of half-curves, where both "halves" are of the same Pearsonian type, but with different parameters.

For $|c| < 1$, the most interesting curves defined by this equation are

the "double" Gaussian $f(x) \propto \exp\{-c_0(x - a)^2/2\sigma^2\}$ \hfill (1.19)

and

the "double" exponential $f(x) = \gamma^{-1}\exp\{-c_0|x|/2\gamma\}$. \hfill (1.20)

The latter curve corresponds to the difference between two exponential variates with different scale parameters and, as such, has considerable intuitive appeal. The curve is further discussed in Exercise 1.5 and a plot of its β_1, β_2 values appears in Fig. 1.1. The estimation of a discontinuity point is discussed by Chernoff and Rubin (1955).

10

1.6 Fitting the curves

Having characterised the curves, where possible, by the first four moments, Pearson proposed the use of the sample moments as estimators of the population moments, to fit a curve to data via (1.8). This enabled a unique choice to be made from among the non-truncated curves satisfying the differential equation (1.4). Using the β_1, β_2 coefficients, Pearson constructed a chart enabling one to select the appropriate type, given the (sample) values of β_1 and β_2. This chart is reproduced as Fig. 1.1.

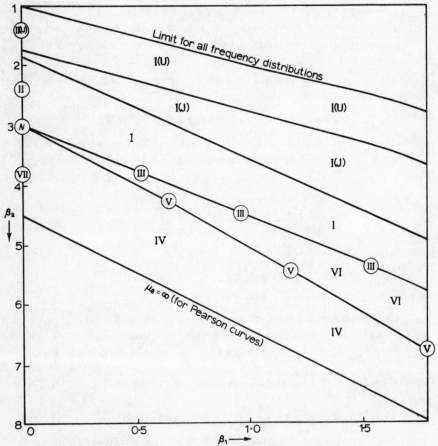

Fig. 1.1 The β_1, β_2 chart for the Pearson curves

Equations of bounding curves

Upper limit for all frequency distributions: $\beta_2 - \beta_1 - 1 = 0$.
Boundary of $I(J)$ area: $4(4\beta_2 - 3\beta_1)(5\beta_2 - 6\beta_1 - 9)^2 = \beta_1(\beta_2 + 3)^2(8\beta_2 - 9\beta_1 - 12)$.
Type III line: $2\beta_2 - 3\beta_1 - 6 = 0$.
Type V line: $\beta_1(\beta_2 + 3)^2 = 4(4\beta_2 - 3\beta_1)(2\beta_2 - 3\beta_1 - 6)$.

(Based on Table 48 of *Biometrika Tables for Statisticians*, Volume I, 1963)

When the moments are known from theoretical considerations this method is acceptable, particularly in those cases where one requires an approximation for evaluating the tail regions of a distribution (see Exercise 1.2). However, fitting a curve to data is quite a different matter.

Karl Pearson (1936) and Fisher (1936) clashed violently on methods of curve fitting; the dispute was sparked off by a paper by Koshal (1933), in which he fitted a beta distribution by maximum likelihood rather than by the method of moments.

With the advantages of hindsight and subsequent developments in statistical theory we can better understand the points at issue.

(a) Fisher's contention that the method of moments was inefficient for Pearson curves, except those near the normal, is now universally accepted. Fisher showed that moments were efficient only for distributions of the form

$$f(x) \propto \exp(a_1 x + a_2 x^2 + \ldots + a_n x^n), \tag{1.21}$$

which is approximately of Pearson form only if a_i, $i > 2$ are small, i.e. near the normal.

In fact, the asymptotic efficiency (ratio of variances) for fitting by moments exceeds 80 per cent only for $\beta_1 < 0\cdot1$, $2\cdot62 \leqslant \beta_2 \leqslant 3\cdot42$, when the first four moments are used.

The lack of non-asymptotic results would perhaps make this difficult to accept when it first appeared.

(b) No grouping corrections were made for the ML method. Maximum likelihood estimation is not necessarily consistent for grouped data (Lindley, 1950), and although this was not explicitly recognised at the time, the need for moment grouping corrections was. While the resulting estimators would still be inconsistent, Pearson's use of "abruptness coefficients" (Pairman and Pearson, 1919) might improve the fit of the curve.

Consistency may not be a very relevant property, but the possibility of correcting the estimates for grouping effects deserved investigation.

(c) The chi-square goodness-of-fit statistic was used to compare the methods. Pearson observed that ML estimation did not lead, necessarily, to a minimum value of χ^2 (although the ML method and minimum χ^2 method have asymptotically the same efficiency). Thus, judging by the computed chi-square values, the moments fit might well come out better, although the value of the likelihood function would be lower.

Unfortunately, even this judgement is not really valid as the computed value is only asymptotically distributed as a chi-square variable if the estimators are efficient. Also, it would appear that Koshal used the data, via Pearson's β_1, β_2 chart, to select the beta distribution, thereby invalidating a subsequent test of fit.

The estimation of parameters from grouped data has attracted attention in recent years and it is possible to obtain consistent estimators. The obstacles in (c), however, remain. If we "dredge" the data for a suitable model, any subsequent test of fit is invalidated (Selvin and Stuart, 1966). Alternatively, if we regard the Pearson system as a family of curves indexed by a parameter, τ say, we must first estimate τ, using an estimator that is efficient for grouped data. Given this estimate, we may select and fit an appropriate curve, again using estimators which are efficient for the grouped data. A subsequent goodness-of-fit test, using chi-square, should allow a degree of freedom for the estimation of τ even if the curve selected gave a particular value for the parameter. This is a slightly conservative test procedure, but not markedly so unless the number of classes is small (cf. section 6.9).

The κ criterion could be used to distinguish the different types, but its estimation via β_1 and β_2 is clearly inefficient. The w criterion defined in section 1.7 could also be used, but again, the moments estimator will not be efficient.

The general problem of discriminating between quite distinct models is of considerable theoretical and practical importance, but progress to date has been limited (see Atkinson, 1970).

In the present context the search for a suitable discriminator τ, which may be efficiently and conveniently estimated, is probably futile. Provided the numbers of observations and classes are not too small, use of κ or w, followed by the use of grouped maximum-likelihood estimators, should not unduly influence a goodness-of-fit test using the usual χ^2 criterion; but this is clearly a matter needing further investigation.

1.7 The *I*, *S* chart

If we know from prior considerations — or are prepared to assume — that the curve has at least one finite end-point, we can proceed as follows.

Take the finite end-point to be at the lower end of the range (without loss of generality, as we may change signs). If the lower end-point, u, is unknown, it may be estimated by a function of the order statistic $x_{(1)}$. Following Robson and Whitlock (1964), we could use the estimator

$$\tilde{u} = 2x_{(1)} - x_{(2)}, \qquad (1.22)$$

for u, which is generally unbiased to $O(n^{-2})$, for a sample of size n.

As in Ord (1967b), we may use the ratios $I = \mu_2/\mu_1'$, $S = \mu_3/\mu_2$ to distinguish the curves. We use the criterion $w = S/2I$ as a single criterion of selection.

Example 1.1

The Gamma (Type III) distribution with density

$$f(x) \propto (x/\gamma)^{k-1} e^{-x/\gamma}$$

has $\mu_1' = k\gamma \qquad \mu_j = (j-1)! \, k\gamma^j,$

whence $S = 2I = 2\gamma$ and $w = 1.$

We may compare the w, κ criteria diagramatically; the following κ diagram is taken from Elderton and Johnson (1969).

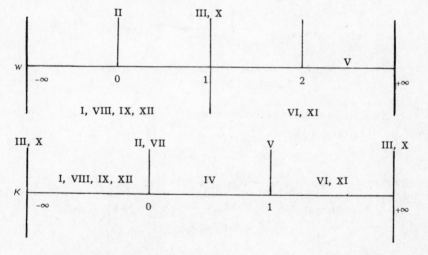

The I, S results for the system are presented in graphic form in Fig. 1.2.

Although the w criterion is more restricted in scope, it leads to simpler calculations and is a more efficient discriminator where applicable; in fact, when the curve starts at $x = 0$, the Pearson equation reduces to

$$\frac{d \ln f}{dx} = \frac{1 - bx}{x(ax + c)}, \quad 0 \leqslant x \leqslant v \leqslant \infty. \tag{1.23}$$

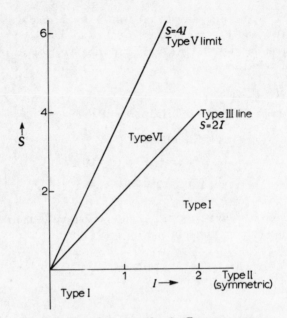

Fig.1.2   The I, S chart for the Pearson curves
(Adapted from Ord, 1967b)

This leads to the moment recurrence relation, provided μ'_{j+1} exists,

$$(aj + 2j - b)\mu'_{j+1} + (cj + c + 1)\mu'_j = v^{j+1}(av + c)f(v)$$
$$= 0 \quad \text{when } v = -c/a \text{ or } f(v) = 0. \tag{1.24}$$

Putting $\alpha_1 = \mu'_1, \quad \alpha_2 = \mu'_2/\mu'_1, \quad \alpha_3 = \mu'_3/\mu'_2$ we find that

$$\left.\begin{aligned}
a &= (2\alpha_2 - \alpha_1 - \alpha_3)/(3\alpha_2 + \alpha_2\alpha_3 - 4\alpha_1\alpha_3) \\
c &= (a\alpha_1\alpha_2 + \alpha_1 - \alpha_2)/(\alpha_2 - 2\alpha_1) \\
b &= \{a(3\alpha_2 - 4\alpha_1) - 1\}/(\alpha_2 - 2\alpha_1),
\end{aligned}\right\} \tag{1.25}$$

which may be used for fitting as previously.

While there are many methods of estimation other than moments and maximum likelihood, we discuss two other methods using
   (a) quantiles and order statistics,
   (b) frequency moments.

1.8 Quantiles and order statistics

(a) *Quantiles*

We proceed by equating the observed $(100p)$th percentile $x(p)$ to its theoretical value $\xi(p)$, for several p, and then solve for the parameters.

Example 1.2

For the normal,

$$x(0{\cdot}5) = \mu \qquad (1.26)$$

$$x(0{\cdot}75) - x(0{\cdot}25) = 1{\cdot}349\sigma \qquad (1.27)$$

and these percentiles have asymptotic efficiencies of 64 per cent, 37 per cent for μ, σ respectively.

(b) *Order statistics*

We equate the jth observed order statistic $x_{(j)}$ to its expected value $E(x_{(j)})$, for appropriate values of j. Corresponding to Example 1.2, we would choose the median and interquartile range.

The two approaches are basically similar, although the results may differ for finite samples. The choice depends primarily on which method is the more tractable mathematically. It is often useful to combine this method with others — we might, for example, use the mean and (1.27) to estimate the normal parameters. In more complex cases, we could use quantile estimates as the first step of an iterative procedure to find the ML estimates.

The quantiles (and order statistics) offer a method of estimation which is widely applicable, often tractable, and reasonably efficient; this approach also enables us to estimate the parameters of truncated distributions (e.g. for life-testing). The main drawback is the standard one — that reasonable efficiency depends on a careful choice of quantiles, and full efficiency is rarely obtained.

For the Pearson system we may use a mixed moment—quantile method. Using the mean and standard deviation, express the random variable in standard measure; then, from the tables of Pearson curve percentage points (Johnson *et al.*, 1963 — see Appendix B for details) for given $\sqrt{\beta_1}$, β_2, either

(a) take two or more sample percentiles and search the tables for compatible $\sqrt{\beta_1}$, β_2 values, or
(b) compute $\sqrt{\beta_1}$ and then find β_2 values compatible with one or more percentiles. This could be used as a starting-point for (a).

If several percentiles are used, we could choose "best" values by minimising some sum of squared deviations. Given these values, we can obtain estimates for the parameters from (1.8) or (1.25).

The efficiency of this approach is unknown, but likely to be an improvement on the four-moment method for an appropriate choice of sample percentiles.

1.9 Location estimators

For some Pearson curves (e.g. the Cauchy), the first moment fails to give a reasonable, or even any, estimate of the location parameter. The symmetric Pearson curves are Types II, VII, the normal and, allowing a single discontinuity, the double exponential. For a slightly different class — excluding Type II, except for the uniform, but including the triangular and the finite mixture of two normals (see section 4.5) — Crow and Siddiqui (1967) have investigated robust estimation of the location parameter. This was followed up by Siddiqui and Raghunandanam (1967). They consider the following means; x_j is here the jth order statistic and $p = j/n$.

Winsorised (Tukey, 1962):

$$W_{.n}(p) = n^{-1}\{(j + 1)(x_{j+1} + x_{n-j}) + \sum_{i=j+2}^{n-j-1} x_i\}, \qquad (1.28)$$

$$j < (n - 1)/2$$

$$W_n(1/2n) = x_{m+1} \quad \text{if } n = 2m + 1.$$

Trimmed (Tukey, 1962):

$$T_n(p) = (n - 2j)^{-1} \sum_{i=j+1}^{n-j} x_i. \qquad (1.29)$$

Linearly weighted:

$$L_n(p) = k_n\left\{\sum_{i=0}^{m}(2i - 2j - 1)(x_i + x_{n-i+1}) + \delta(2m - 2j + 1)x_{m+1}\right\} \qquad (1.30)$$

where $\quad k_n^{-1} = 2(m - j)^2 + \delta(2m - 2j + 1)$,

and $\qquad \delta = 1, \quad \text{if } n = 2m + 1$

$\qquad\qquad = 0, \quad \text{if } n = 2m.$

Median plus two other symmetric order statistics:

$$Y_n(p, a) = a(x_{j+1} + x_{n-j}) + (1 - 2a)M \qquad (1.31)$$

where $M = x_{m+1}$ if $n = 2m + 1$

$\quad\quad\quad = \frac{1}{2}(x_m + x_{m+1})$ if $n = 2m$.

From consideration of various small-sample and asymptotic results, they suggest using either a trimmed mean with $j \simeq n/4$ or a linearly weighted mean with $j \simeq n/6$ or a median plus order-statistics estimator with $j \simeq n/4$ and $a = 0 \cdot 25$.

For the distributions mentioned, except the uniform and Type II, asymptotic efficiencies are in the region of 80 per cent or more using these methods, without any strong evidence emerging for any one rather than the other two.

Birnbaum and Laska (1967) consider the family formed by mixing the normal (N), logistic (L), and double exponential distributions in the proportions $\{p_\lambda,\ \lambda = N, L, D: \Sigma p_\lambda = 1\}$; see section 4.1. The weights for the weighted linear estimator (WLE) are chosen to minimise

$$\max_{\{p_\lambda\}} \left\{ \frac{\text{var (WLE)}}{\text{var (ML estimator)}} \right\}.$$

More generally, E.S. Pearson and Tukey (1965) have suggested the estimator

$$Y_n(0 \cdot 05, 0 \cdot 185) = x(0 \cdot 50) + 0 \cdot 185 \Delta,$$

where $\Delta = x(0 \cdot 95) + x(0 \cdot 05) - 2x(0 \cdot 50)$, for the Pearson distributions, skewed as well as symmetric. Their empirical findings suggest that the large-sample bias is small for a wide variety of curves. They also give a robust estimator for the standard deviation as $\hat{\sigma}$, the solution of the equations

$$\hat{\sigma} = \max \{\hat{\sigma}(0 \cdot 05),\ \hat{\sigma}(0 \cdot 025)\},$$

$$\hat{\sigma}(0 \cdot 05) = \frac{\{x(0 \cdot 95) - x(0 \cdot 05)\}}{\max [3 \cdot 29 - 0 \cdot 1 \{\Delta/\hat{\sigma}(0 \cdot 05)\}^2,\ 3 \cdot 08]},$$

and $\hat{\sigma}(0 \cdot 025) = \dfrac{\{x(0 \cdot 975) - x(0 \cdot 025)\}}{\max [3 \cdot 98 - 0 \cdot 38 \{\Delta/\hat{\sigma}(0 \cdot 025)\}^2,\ 3 \cdot 66]}.$

1.10   Frequency moments

Yule (1938) pursued a suggestion by Karl Pearson that instead of using the moments one might use the frequency moments

$$\omega_j = \int \{f(x)\}^j dx, \quad \omega_1 \equiv 1. \tag{1.32}$$

If the value of the density function at the mode, $f(m)$, is finite, clearly $\omega_j < \{f(m)\}^{j-1}$ exists for all j; in particular this is so for the Pearson Type IV and Type VII curves. If the maximum value of $f(x)$ is unbounded, only a finite number of ω_j may exist; for example, when

$$f(x) \propto x^{-a}e^{-x}, \quad 0 < a < 1, \quad 0 \leqslant x < \infty,$$

the ω_j exist only while $ja < 1$.

The ω_j lack a location parameter — two curves differing only in location have identical frequency moments — but this could be provided by the median or one of the functions discussed in section 1.9.

Sichel (1947, 1949) developed the theory of these estimators, obtaining the following sampling results for the observed frequency moments a_j, for sample size n:

$$E(a_j) = \omega_j + \frac{j(j-1)}{n}(\omega_{j-1} - \omega_j) + O(n^{-2}) \tag{1.33}$$

$$\text{var}(a_j) = j^2/n \, (\omega_{2j-1} - \omega_j^2) + O(n^{-2}). \tag{1.34}$$

These results hold asymptotically for all $j > 1$, but the most useful values of j are $k/2 + 1$, $k = 1, 2, 3, \ldots$.

More particularly, Sichel establishes that

$$\text{var}(a_2) = 4n^{-1}(\omega_3 - \omega_2^2) + 2n^{-2}(\omega_2 - 6\omega_3 + 5\omega_2^2) + \\ + 2n^{-3}(\omega_2 - 4\omega_3 + 3\omega_2^2), \tag{1.35}$$

while

$$\beta_1(a_2) = \frac{(4\omega_4 + 5\omega_2^3 - 9\omega_2\omega_3)^2}{n(\omega_3 - \omega_2^2)^3} + O(n^{-2}), \tag{1.36}$$

and

$$\beta_2(a_2) = 3 + O(n^{-2});$$

however, the convergence to normality is slow.

Estimation from grouped data requires weights to be attached to the end groups; these are, for k groups,

$$
\begin{aligned}
w_1 &= w_k &= 1 \cdot 122 \, 05 \\
w_2 &= w_{k-1} &= 0 \cdot 758 \, 85 \\
w_3 &= w_{k-2} &= 1 \cdot 157 \, 81 \\
w_4 &= w_{k-3} &= 0 \cdot 961 \, 29 \\
w_i &= 1, \ i = 5, \ldots, k-4.
\end{aligned}
$$

This awkwardness in application, coupled with a lack of either theoretical results or obvious interpretations, explain why the method is so little used. However, for distributions where the moment solution either does not exist or gives very poor results, especially in the Pearson heterotypic region, the frequency moments often yield high efficiency.

Example 1.3

The Type VII distribution has density function

$$f(x) = \{(1 + x^2/b^2)^k \, b\beta(\tfrac{1}{2}, k - \tfrac{1}{2})\}^{-1} \tag{1.37}$$

with frequency moments, found directly by integration,

$$\omega_j = b^{1-j} \, \beta(\tfrac{1}{2}, jk - \tfrac{1}{2}) \, \{\beta(\tfrac{1}{2}, k - \tfrac{1}{2})\}^{-j}. \tag{1.38}$$

Further, for the Cauchy, the log-likelihood function for a sample of size n is

$$\ln L \propto n \ln b - \sum_{i=1}^{n} \ln (b^2 + x_i^2),$$

whence $\quad \dfrac{\partial \ln L}{\partial b} = n/b - \sum \dfrac{2b}{b^2 + x_i^2}.$

Taking expectations,

$$E\left(\frac{\partial \ln L}{\partial b}\right) = n(b^{-1} - 2\pi\omega_2), \tag{1.39}$$

showing that the frequency moment is asymptotically equivalent to the ML estimator; both estimators having asymptotic variance $2b^2/n$.

It may be shown (Ord, 1968b) that if the density function obeys the differential equation

$$\frac{df(x)}{db} = fh(g - f) \tag{1.40}$$

where h, g are functions of b only, and the range of x does not depend on b, then ω_2 is asymptotically fully efficient for b. The solution to (1.40) has the general form

$$f(x) = \mu/\{\int h\mu \, db + c(x)\}, \tag{1.41}$$

where $\mu = \exp(\int hg \, db)$, and c is a function of x only. A particular solution is

$$f(x) \propto \{b^j + x^j\}^{-1}, \quad j > 1. \tag{1.42}$$

Provided the set of possible values for b is continuous, so that (1.40) is valid, (1.41) applies equally to continuous and discrete distributions; in particular, ω_2 is asymptotically efficient for both the continuous and the discrete Cauchy (section 5.3).

1.11   Examples using Pearson curves

Example 1.4

The non-central chi-square distribution with k degrees of freedom and non-centrality parameter δ^2 has density function, $x = \frac{1}{2}\chi^2$,

$$f(x) = e^{-x-\delta^2} \sum_{m=0}^{\infty} \delta^{2m} x^{m+1-k/2} / \{m! \, \Gamma(m + k/2)\}.$$

Its cumulants are $\kappa_j = (j - 1)! \, (k + j\delta^2), \quad j = 1, 2, \ldots,$

yielding $w = S/2I = -2\delta^2/(k + \delta^2),$

which satisfies $-2 \leqslant w \leqslant 0$, suggesting that we use the Type III (central χ^2) for small δ^2 and Type I for larger δ^2 as approximations.

E.S. Pearson (1963) discussed the use of Type I as an approximation to the non-central χ^2; since the β_1, β_2 values lie between the lines

$$2\beta_2 - 3\beta_1 - 6 = 0 \quad \text{(Type III line)}$$

and $3\beta_2 - 4\beta_1 = 9 = 0$

the same conclusions as above hold. To improve the fit in the upper tail he also allowed the location of the start of the curve to vary.

Example 1.5

From Elderton and Johnson (1969, p. 79) we take Table 1.2 giving for wives in the 30–34 age range who were mothers, the number of years (x) between marriage and arrival of the first child. The data record for x starts at $x = 0 \cdot 5839$ or 7 months after marriage.

Table 1.2

Year after marriage	Number of wives	Type III fit by moments	Type X fit
1	44	59	56
2	135	111	107
3	45	45	48
4	12	20	22
5	8	9	10
6	3	4	4
7	1	2	2
8	3	1	2

Elderton and Johnson give the sample moments

$$m_1' = 1\cdot8335 \qquad m_2 = 1\cdot4418$$
$$m_3 = 3\cdot6066 \qquad m_4 = 18\cdot9322$$

yielding $w = 1\cdot08$, $\kappa = -8\cdot44$. The fitted Type III curve starts at $x = 0\cdot6819$ with index $0\cdot9216$ and scale parameter $\gamma = 1\cdot2507$.

The sample values of β_1, β_2 are $4\cdot3$, $9\cdot1$ respectively, close to those $(4, 9)$ of the exponential or Type X. This curve has considerable theoretical attraction for the above data as it corresponds to the assumption of "random arrivals". Table 1.2 shows the fit for the Type X, starting at $x = 0\cdot5839$ and using scale parameter $\gamma = 1\cdot2596$ obtained from m_1'.

Since three parameters are fitted in the first case and only one in the second, the Type X model seems as good as the Type III — though neither fit is particularly impressive.

1.12 The exponential family

We mentioned in section 1.6 that the moment estimators were fully efficient for $f(x) \propto \exp(a_1 x + a_2 x^2 + \dots + a_k x^k)$, as shown by Fisher (1921a). This form is of little interest for general k, but when only $a_1 (= \theta)$ is unknown, we have the exponential family with

$$\ln f(x, \theta) = x\theta + A(x) + B(\theta), \qquad (1.43)$$

or more generally

$$\ln f(x, \theta) = t(x) q(\theta) + A(x) + B(\theta),$$

where t and q are monotone functions in x, θ respectively, and $\theta \in \Omega$, where $\Omega = \{\theta \,|\, B(\theta) < \infty\}$.

For a sample of size n, the likelihood is maximised when

$$t_\sigma = \sum_{i=1}^{n} t(x_i) = -B'(\theta)/q'(\theta), \qquad (1.44)$$

unless the range of x depends on θ; thus t_σ is sufficient for θ. This class is of considerable theoretical importance and includes the normal and Type III curves; for a discussion of the family in the discrete case see section 6.2 *et seq.* on the power series distributions.

We now present several properties of the family.

Property E1. Since $\int_{-\infty}^{\infty} \exp\{x\theta + A(x)\}\,dx = \exp\{-B(\theta)\}$, it follows that the characteristic function, $\phi(t)$, is

$$\phi(t) = \exp\{B(\theta) - B(\theta + it)\}. \qquad (1.45)$$

Property E2. Provided the jth cumulant, κ_j, exists, we may establish, on differentiating $\ln \phi(t)$, that

$$\kappa_j = \frac{d\kappa_{j-1}}{d\theta}, \quad j = 2, 3, \ldots. \tag{1.46}$$

This is both a necessary and sufficient condition for the distribution to be a member of the exponential family — see Patil (1963a).

Property E3. For testing the hypothesis $H_0 : \theta = \theta_0$ against the alternative $H_1 : \theta > \theta_0$, application of the Neyman—Pearson lemma shows the existence of a uniformly most powerful (UMP) test at level α, given by "reject H_0 with probability $\psi(x)$" where

$$\left.\begin{aligned}
\psi(x) &= 1 \text{ when } t_\sigma > c \\
&= \gamma \text{ when } t_\sigma = c \\
&= 0 \text{ when } t_\sigma < c
\end{aligned}\right\} \tag{1.47}$$

c and γ being chosen so that $E\{\psi(x) \mid \theta_0\} = \alpha$. The power function $\beta(\theta) = E\{\psi(x) \mid \theta\}$ is monotone increasing in $\delta = \theta - \theta_0$ for all θ for which $\beta(\theta) < 1$, so that the test is also UMP for $H_0 : \theta \leqslant \theta_0$ against $H_1 : \theta > \theta_1 \geqslant \theta_0$.

For tests of $H_0 : \theta = \theta_0$ against $H_1 : \theta \neq \theta_0$ and $H_0 : \theta_1 \leqslant \theta \leqslant \theta_2$ against $H_1 : \theta > \theta_2$ or $\theta < \theta_1$, the test based on (1.47) is UMP unbiased; that is, the test maximises power subject to

$$E(\psi(x) \mid \theta_1 \leqslant \theta \leqslant \theta_2) \leqslant \alpha$$

and $\quad\quad E(\psi(x) \mid \theta > \theta_2 \quad \text{or} \quad \theta < \theta_1) \geqslant \alpha.$

A full discussion of hypothesis testing for exponential families in both single- and multi-parameter cases is given in Lehmann (1959), including proofs of the above results.

Property E4 (Patil and Shorrock, 1965). The mean of any distribution in the family is $\mu(\theta) = -B'(\theta)$, from (1.45). If $\mu(\theta)$ is given for $\theta \in \Omega$, then the type of distribution is determined uniquely within the exponential family.

This follows from the uniqueness of the characteristic function and (1.45). As an application of this property, if $t(x)$ is the ML estimator for θ and $t(x)$ is given for a sequence $\{x_n\}$ with limit point x_0, the distribution is determined within the family of all exponential type distributions.

Property E5 (Kendall and Stuart, 1967, pp. 24—27). The parametric function $\tau(\theta)$ has a minimum variance bound unbiased (MVBU) estimator if and only if

$$\partial \ln L(\theta)/\partial\theta = c(\theta)\{t(x) - \tau(\theta)\},$$

which for the exponential family implies that

$$\partial \ln L(\theta)/\partial\theta = x_1 + x_2 + \ldots + x_n + n\,B'(\theta). \qquad (1.48)$$

Thus $\tau(\theta)$ has a MVBU estimator if and only if τ is a linear function of μ, i.e. $\tau(\theta) = a\mu(\theta) + b$, a, b constants. Patil and Shorrock (1965) show that this requirement is satisfied only if the first two Bhattacharya bounds are equal on some open interval in the parameter space.

For a further discussion of the exponential family and its properties see Kosambi (1949), Lehmann (1959), Patil and Shorrock (1965) and Doss (1969); and for the power series distributions see section 6.2 and following sections.

EXERCISES

1.1 If the density function satisfies the Pearson differential equation (1.4), use (1.12) to derive a differential equation for the cumulant generating function. Hence, assuming the jth cumulant, κ_j, exists, prove that

$$\{1 + (j+1)b_2\}\kappa_j + (j-1)b_1\,\kappa_{j-1} + (j-1)b_2 \sum_{s=1}^{j-3}\binom{j-2}{s}\kappa_{s+1}\kappa_{j-1-s} = 0.$$

(Kendall, 1941)

1.2 For the standard normal distribution with density

$$a(x) \propto \exp\left(-\tfrac{1}{2}x^2\right)$$

show that

$$\int_{|x|>1.96} x^4\,a(x)dx \simeq 0.82\mu_4.$$

That is, the higher moments of the normal (or any other) curve are heavily dependent on the tail regions. (Mishra, 1955)

1.3 Show that, for the Pearson Type III curve,

$$2\beta_2 - 3\beta_1 - 6 = 0$$

and hence that the κ criterion is infinite. Show also that κ is indeterminate for the normal.

1.4 Instead of using the real random variable x, insert $y = ix$ in the Pearson differential equation. Use the identity $\tan u \equiv i \tanh(iu)$ to establish a relationship between the Pearson Type IV and Type VI distributions. (Kendall and Stuart, 1969, p. 176)

1.5 For the general double exponential distribution (1.20), show that

$$\mu'_1 = -2c\gamma,$$
$$\mu_2 = 2\gamma^2(1 + c^2),$$
$$\beta_1 = 2c^2(3 + c^2)(1 + c^2)^{-3},$$

and
$$\beta_2 = 6 + 12c^2(1 + c^2)^{-2},$$

enabling one to plot the β_1, β_2 values on Fig. 1.1.

1.6 Show that if x follows the double Gaussian distribution (1.19), and y is normally distributed, both having the same mode, the ratio x/y is distributed according to a "double-Cauchy" distribution (defined by analogy), which is a member of family (1.17).

1.7 Find the exact value of $w = S/2I$ for the Type I, Type V and Type VI distributions.

1.8 Use the results of section 1.10 to find the frequency moment estimators for σ, the standard deviation of a normally distributed random variable, based on $\omega_{3/2}$ and ω_2. Show that their asymptotic efficiencies are $9(3\sqrt{2} - 4)/2 = 0{\cdot}916$ and $\sqrt{3}/8(2 - \sqrt{3}) = 0{\cdot}808$ respectively. (Sichel, 1949)

1.9 The generalised Gompertz—Verhulst family of curves comprises three types with distribution functions

$$F_1(x) = \exp\{-c \exp(-x/b)\},$$
$$F_2(x) = \{1 + c \exp(-x/b)\}^{-m},$$

and
$$F_3(x) = \{1 - c \exp(-x/b)\}^m.$$

Show that the variate $y = \exp(-x/b)$ follows the Pearson Types X, VIII and IX respectively. Hence find the characteristic function for each type. (Ahuja and Nash, 1967)

CHAPTER 2

DISTRIBUTIONS BASED ON SERIES EXPANSIONS AND TRANSFORMATIONS OF THE RANDOM VARIABLES

2.1 Series expansions

It was mentioned in section 1.1 that, given some standard density function, $f(x)$, we might express other density functions as series expansions in terms of this standard form. That is, the density function $g(x)$ may be written

$$g(x) = f(x) \sum_{j=0}^{k} c_j \, q_j(x), \qquad (2.1)$$

where the q_j are functions of x, and k may be finite or infinite. We restrict ourselves to the case when the q_j are polynomials in x of order j or less, and obey the orthogonal relations (T being the range of x)

$$\int_T q_i(x) \, q_j(x) \, f(x) \, dx = 0, \quad i \neq j \qquad (2.2a)$$

$$= j!, \quad i = j. \qquad (2.2b)$$

The coefficients of the q_j are determined up to a scale factor by conditions (2.2a), while (2.2b) is a convenient, if arbitrary, norming condition.

From (2.1), on interchanging the sum and integral operations (assuming that $E_f(q_k^2)$, $E_g(q_k)$ exist), we have

$$\int_T q_i(x) \, g(x) \, dx = \sum_{j=0}^{k} c_j \int_T q_i(x) \, q_j(x) \, f(x) \, dx \qquad (2.3)$$

which gives, from (2.2),

$$c_j \, j! = \int_T q_j(x) \, g(x) \, dx. \qquad (2.4)$$

The choice of $f(x)$ is open, and Romanovsky (1924) considered various Pearson curves for the role, yielding for $\{q_j\}$ the classical orthogonal polynomials — see Appendix A. However, the standard normal density function, $\alpha(x)$, has been used in most cases and we will consider only this aspect of the theory.

The Hermite polynomials, which are the orthogonal polynomials for the normal, are discussed in Appendix A, but we give the algebraic

25

26

forms here, for convenience:

$$
\left.
\begin{array}{ll}
H_0 = 1 & H_1 = x \\
H_2 = x^2 - 1 & H_3 = x^3 - x \\
H_4 = x^4 - 6x^2 + 3 & H_5 = x^5 - 10x^3 + 15x \\
H_6 = x^6 - 15x^4 + 45x^2 - 15 &
\end{array}
\right\} \quad (2.5)
$$

The general form $j! \, c_j$ is, from (2.4),

$$
\mu'_j - \frac{j^{(2)} \, \mu'_{j-2}}{2. \, 1!} + \frac{j^{(4)} \, \mu'_{j-4}}{2^2. \, 2!} - \dots , \quad (2.6)
$$

where $\mu'_j = \int_T x^j g(x)\,dx$ and $j^{(r)} = j(j-1)\dots(j-r+1)$. Or, where the origin is taken at the mean,

$$
\left.
\begin{array}{ll}
c_0 = 1 & c_1 = 0 \\
2! \, c_2 = \mu_2 - 1 & 3! \, c_3 = \mu_3 \\
4! \, c_4 = \mu_4 - 6\mu_2 + 3 & 5! \, c_5 = \mu_5 - 10\mu_3 \\
6! \, \mu_6 = \mu_6 - 15\mu_4 + 45\mu_2 - 15. &
\end{array}
\right\} \quad (2.7)
$$

The resulting series,

$$
\begin{aligned}
g(x) = \alpha(x) \{ & 1 + (\mu_2 - 1)H_2/2! + \mu_3 H_3/3! + \\
& + (\mu_4 - 6\mu_2 + 3)H_4/4! + \dots \},
\end{aligned} \quad (2.8)
$$

is known as the Gram–Charlier series Type A (Charlier 1905, 1907).

*For the equality in (2.8) to hold, excepting sets of zero measure, we require all moments of $g(x)$ to exist; while if $g(x)$ were defined on a lattice of width Δx, we would need to assume

$$
g(x) = p(x_0), \quad x_0 - \tfrac{1}{2}\Delta x \leqslant x < x_0 + \tfrac{1}{2}\Delta x \quad (2.9)
$$

and require

$$
x/\Delta x = -m, \dots, -1, 0, 1, \dots, n,
$$

$$
\lim_{m+n \to \infty} \Delta x = 0.
$$

2.2 The Edgeworth series

Instead of adapting the density function, we might start with the characteristic function

$$
\begin{aligned}
\phi_g(t) &= \int_T e^{\theta x} g(x)\,dx, \quad \theta = it \\
&= \exp\left(\sum_{j=1}^{\infty} \kappa_j \theta^j / j! \right),
\end{aligned} \quad (2.10)
$$

κ_j being the jth cumulant of $g(x)$, all of which are assumed to exist. If $f(x)$ has cumulants $\{\eta_j\}$, all of which exist, it has the characteristic function

$$\phi_f(t) = \exp\left(\sum_{j=1}^{\infty} \eta_j \theta^j/j!\right),$$ (2.11)

so we may write

$$\phi_g(t) = \exp\left\{\sum_1^{\infty} (\kappa_j - \eta_j)\theta^j/j!\right\} \phi_f(t).$$ (2.12)

Using the inversion theorem for the Fourier transform,

$$f(x) = \int_S e^{-\theta x} \phi_f(t)\,dt, \quad t \in S$$ (2.13)

and assuming it is valid to differentiate under the integral sign, we have

$$(-D)^j f(x) = \int_S \theta^j e^{-\theta x} \phi_f(t)\,dt, \quad j = 1, 2, \ldots,$$ (2.14)

where $D = d/dx$.

Since all the cumulants exist, (2.12) may be written as

$$\phi_g(t) = \left\{1 + \sum_1^{\infty} a_j \theta^j\right\} \phi_f(t)$$

and (2.13), (2.14) yield

$$g(x) = \left\{1 + \sum_1^{\infty} a_j(-D)^j/j!\right\} f(x).$$ (2.15)

In general, $q_j(x) = \{f(x)\}^{-1}(-D)^j f(x)$ will not be polynomial in form, but it is evident from the Rodrigues relation, (A.4) in Appendix A, that when $f(x)$ is the normal density function $\alpha(x)$, we have

$$(-D)^j \alpha(x) = H_j(x)\,\alpha(x)$$ (2.16)

– the Hermite polynomials again. For the normal, $\eta_1 = \mu$, $\eta_2 = \sigma^2$ and $\eta_j = 0$, $j \geqslant 3$, so we find on comparing coefficients and equating the first two moments, that the first seven a_i are

$$
\begin{aligned}
a_0 &= 1 & a_1 &= a_2 = 0 \\
a_3 &= \kappa_3/6 & a_4 &= \kappa_4/24 \\
a_5 &= \kappa_5/120 & a_6 &= (\kappa_6 + 10\kappa_3^2)/720.
\end{aligned}
\right\}
$$ (2.17)

This expansion was first derived by Edgeworth (1896) from the theory of elementary errors, but the above approach is that given by Kendall and Stuart (1969), pp. 156–61, on which sections 2.1–2.3 are based.

2.3 Existence of series expansions

We have already stated that the cumulants of all orders must exist for both $g(x)$ and $f(x)$; for $f(x)$ normal, Cramér (1926) derived the following limit theorems — proofs are given in Kendall and Stuart (1969).

(1) If $g(x)$ is a function with a continuous derivative, g_x, such that

$$\int_{-\infty}^{\infty} g_x^2\, e^{\frac{1}{2}x^2}\, dx < \infty \tag{2.18}$$

and $\lim_{x \to \infty} g(x) = 0$, then the series

$$g(x) = \alpha(x) \sum_{j=0}^{\infty} c_j H_j(x)/j! \tag{2.19}$$

with c_j given by (2.7) or as in (2.17), is absolutely and uniformly convergent for $-\infty \leqslant x \leqslant \infty$.

(2) If $g(x)$ is of bounded variation in every interval and if

$$\int_{-\infty}^{\infty} |g(x)|\, e^{x^2/4} dx < \infty,$$

then $g(x)$ as defined by (2.19) exists and is uniformly convergent in every finite interval.

Cramér shows that this second theorem is about as far as we can go in establishing the existence of the expansions, which means we must impose rigorous restrictions. Furthermore, even when (2.19) does converge, the tetrachoric series — so called because $\tau_j(x) = H_{j-1}(x) \times \alpha(x)(j!)^{-\frac{1}{2}}$ were labelled tetrachoric functions — may do so only very slowly, as shown by Henderson (1925).

When x depends on a parameter $n\ (> 1)$, e.g. sample size, Feller (1966, pp. 506–9) gives an elegant proof of the following

Theorem

Suppose that the moments $\mu_3, \dots, \mu_s$ exist and that $|\phi(\theta)|^\nu$, where $\phi(\theta)$ is the characteristic function, is integrable for some $\nu \geqslant 1$. Then $g_n(x)$ given by (2.15) exists for $n \geqslant \nu$, and, as $n \to \infty$,

$$g_n(x) - \alpha(x) - \alpha(x) \sum_{j=3}^{s} n^{-\frac{1}{2}j+1} H_j(x) = o(n^{-\frac{1}{2}s+1}), \tag{2.20}$$

$H_j(x)$ being the Hermite polynomials.

2.4 Statistical uses

Edgeworth believed his series to be different from the Gram–Charlier form, but it is clear from the uniqueness of the characteristic function that if both series exist they must give the same values for $g(x)$. This line of reasoning led some authors (e.g. Carver, 1919) to conclude that both the series and Pearson approaches must ultimately yield the same results as taking the Pearson equation in the general form

$$\frac{d \ln f}{dx} = \frac{x - a}{U(x)}. \tag{2.21}$$

This is true provided that $U(x)$ does not have to be expressed as a polynomial to be convergent for all x, and the series expansions are complete. However, for statistical uses, we generally truncate the series forms after only three or four terms, while $U(x)$ is restricted to being a quadratic; it is this truncation which gives rise to differences between the approaches.

From a data-fitting viewpoint it is essential to use only a few terms because of large sampling fluctuations in the higher coefficients, while, for theoretical uses, terms as far as κ_6 usually suffice, as higher terms increase the risk of negative probabilities in the tails without appreciably improving an approximation.

Rather than consider ultimate convergence, therefore, we look at the following features:

(a) size of errors through neglect of higher order terms,

(b) whether the series for $g(x)$ is everywhere non-negative,

(c) whether the series is unimodal or multimodal.

2.5 Size of errors

Boas (1949) showed that the Gram–Charlier Type A series is optimal in the sense that

$$\int_{-\infty}^{\infty} \left\{ g(x) - \sum_{j=0}^{s} c_j H_j(x) \, a(x) \right\}^2 \{a(x)\}^{-1} \, dx \tag{2.22}$$

is minimal for all s, when the $\{c_j\}$ are given by (2.7).

If $g(x)$ is the distribution of the sample mean, from a sample of size n, with cumulants κ_1, $\kappa_2 = \sigma^2$, κ_3, ... , we can express these cumulants in standard measure as $l_j = \kappa_j \sigma^{-j}$, where l_j is $O(n^{1-j/2})$. Because of the method of generation, the Edgeworth series can be regrouped to give

$$g(x) = \alpha(x) \left[1 + l_3 H_3(x)/6 + \{3l_4 H_4(x) + l_3^2 H_6(x)\}/72 + \right.$$
$$\left. + \{6l_5 H_5(x) + 5l_3 l_4 H_7(x)\}/720 + ...\right] \qquad (2.23)$$

where successive terms are of decreasing order in n. The Gram–Charlier Type A could also be modified to give (2.23) but only by introducing additional terms; without the modification, the order in n of successive terms does not decrease regularly, so the asymptotic properties of the series must be investigated very carefully.

Without re-ordering, the order of the coefficients in the Type A series is

$$c_{3j+s} \sim O(n^{-(j+s)/2}) \qquad s = 0, 1, 2, ... ; \qquad j = 1, 2, ...$$

The above analysis applies for any parameter n for which l_j is $O(n^{1-j/2})$, not just sample size.

2.6 Modality and non-negativity

Barton and Dennis (1952) investigated the truncated forms

Gram–Charlier:

$$g(x) = \alpha(x) \{1 + \kappa_3 H_3(x)/6 + \kappa_4 H_4(x)/24\} \qquad (2.24)$$

Edgeworth:

$$g(x) = \alpha(x) \left[1 + \kappa_3 H_3(x)/6 + \{3\kappa_4 H_4(x) + \kappa_3^2 H_6(x)\}/72\right] \qquad (2.25)$$

to see for which values of the Pearson coefficients β_1, β_2 they (a) were unimodal and (b) had $g(x)$ non-negative for all x.

The functions (2.24) and (2.25) are in fact the most commonly used forms of the expansions, since truncation after two terms clearly leads to negative values in one tail for $g(x)$, though this may only occur for extreme x. Their results are presented in Fig. 2.1, and it is worth noting that these expansions are of value mainly in the region occupied by Pearson's Type IV. The latter being generally regarded as a mathematical accident rather than a statistical blessing, these series forms are seen to be complementary to the Pearson system rather than in competition with it.

The saddlepoint approximations of Daniels (1954) are a generalisation of Edgeworth's method, with the advantage of yielding a non-negative density function always.

31

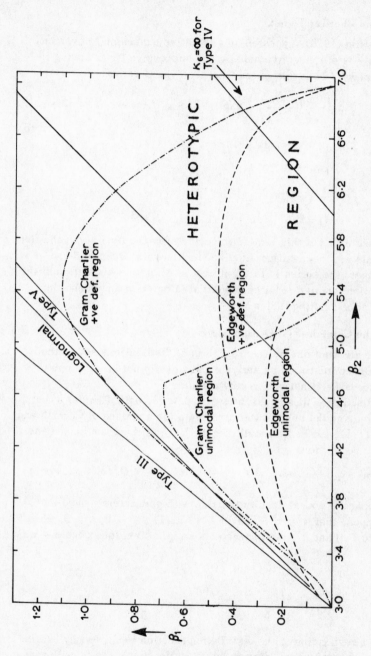

Fig. 2.1 The unimodal and positive definite regions in the β_1, β_2 plane for the Edgeworth and Gram–Charlier series (The Type IV region is the whole area to the right of the Type V curve.) (Adapted from Barton and Dennis, 1952)

2.7 The Charlier Type C

Charlier (1928) suggested an alternative expansion to overcome the problems of a negative density and successive terms having irregular powers in n; this is the Charlier Type C series:

$$g(x) = \exp\left\{-\frac{1}{2}\left(\frac{x-\mu}{\sigma}\right)^2\right\} \exp\left\{-c_0 - c_1 H_1(x) - c_2 H_2(x)/2 - \ldots\right\}$$
(2.26)

where $c_0 = (4l_3^2 + l_4^2)/48 + O(n^{-2}) \sim O(n^{-1})$

$c_1 \sim O(n^{-3/2})$

$c_2 = \frac{1}{2}l_3^2 + O(n^{-2}) \sim O(n^{-1})$

$c_j = l_j - O(n^{-j/2}), \sim O(n^{1-j/2}), \quad j \geqslant 3.$

We note that if this form is truncated after the term in c_j, the first j moments are the maximum likelihood estimators. Despite these advantages, the series is rarely if ever used, as the evaluation of the distribution function requires numerical integration and cannot be derived from existing tables.

2.8 The Cornish–Fisher expansion

The dividing line between systems of distributions and methods of approximation based on series expansions is not an easy one to draw, nor is its construction very useful.

A discussion of approximations is deferred until Chapter 8, but before leaving the subject we investigate the approach of Cornish and Fisher (1937, 1960). We recall that the Edgeworth expansion may be written

$$g(x) = \exp\left\{-(\kappa_1 - \eta_1)D + (\kappa_2 - \eta_2)D^2/2! - \kappa_3 D^3/3! + \ldots\right\} a(x),$$
(2.27)

$a(x)$ being the normal density function with parameters η_1, $\eta_2 (= \sigma^2)$.

Suppose that $\kappa_i - \eta_i = l_i \sigma^i$, $i = 1, 2, \ldots$; $\eta_i = 0$, $i \geqslant 3$; when κ_i is $O(n^{1-i})$ and n some parameter – sample size, for example – we have

$$l_1, l_3 \quad \text{are} \quad O(n^{-\frac{1}{2}})$$

$$l_2, l_4 \quad \text{are} \quad O(n^{-1})$$

$$l_i \quad \text{is} \quad O(n^{1-i/2}), \quad i \geqslant 5.$$

Thus, corresponding to the distribution function $G(x)$, we may obtain a series in terms of the Hermite polynomials, $H_i(x)$, and l_i, with terms grouped according to their power in n.

So, if $\Phi(x)$ is the normal distribution function, we find that

$$G(x) = \Phi(x) - \alpha(x) \sum_{j=1}^{\infty} T_j \qquad (2.28)$$

where T_j is $O(n^{-j/2})$. The first few T_j are given below, where $H_j \equiv H_j(x)$.

$$T_1 = l_1 + l_3 H_2/6; \quad T_2 = \frac{1}{2}(l_1^2 + l_2)H_1 + \frac{1}{24}(4l_1 l_3 + l_4)H_3 + \frac{1}{72}l_3^2 H_5;$$

$$T_3 = \frac{1}{6}(l_1^3 + 3l_1 l_2)H_2 + \frac{1}{120}(10l_1^2 l_3 + 10l_2 l_3 + 5l_1 l_4 + l_5)H_4 +$$

$$+ \frac{1}{144}(2l_1 l_3^2 + l_3 l_4)H_6 + \frac{1}{1296}l_3^3 H_8;$$

$$T_4 = \frac{1}{24}(l_1^4 + 3l_2^2 + 6l_1^2 l_2)H_3 + \frac{1}{720}(20l_1^3 l_3 + 60l_1 l_2 l_3 + 15l_1^2 l_4 + 15l_2 l_4 +$$

$$+ 6l_1 l_5 + l_6)H_5 + \frac{1}{5760}(40l_1^2 l_3^2 + 40l_2 l_3^2 + 5l_4^2 + 40l_1 l_3 l_4 + 8l_3 l_5)H_7 +$$

$$+ \frac{1}{5184}(4l_1 l_3^3 + 3l_3^2 l_4)H_9 + \frac{1}{31\,104}l_3^4 H_{11}.$$

When $g(x)$ is continuous, $G(x)$ is a monotone increasing function and there exists a one-to-one relation between x and y values, so we could write

$$x - y = a_0 + a_1 x + a_2 x^2 + ..., \qquad (2.29)$$

where $\quad G(x) = \int_{-\infty}^{x} g(x)\,dx = \int_{-\infty}^{y} \alpha(y)\,dy = \Phi(y).$ $\qquad (2.30)$

If we expand $\Phi(y)$ in (2.30) using the Taylor series, then

$$\Phi(x + y - x) = \Phi(x) - \alpha(x) \sum_{j=1}^{\infty} (x - y)^j H_{j-1}(x)/j! \qquad (2.31)$$

as $\quad (-D)^j \Phi(x) = (-D)^{j-1} \alpha(x) = H_{j-1}(x)\,\alpha(x);$

applying (2.29) to (2.31), we obtain the a_i by comparing coefficients. The expansion is then grouped according to powers of n, yielding

$$x - y = \left\{ l_1 + \frac{1}{6} l_3(x^2 - 1) \right\} + \left\{ \frac{1}{6}(3l_2 - 2l_1 l_3)x + \frac{1}{24} l_4(x^3 - 3x) - \right.$$

$$\left. -\frac{1}{36} l_3^2(4x^3 - 7x) \right\} + \left\{ \frac{1}{6} l_1^2 l_3 - \frac{1}{2} l_1 l_2 + \frac{1}{120} l_5(x^4 - 6x^2 + 3) - \right.$$

$$-\frac{1}{12} l_2 l_3(5x^2 - 3) - \frac{1}{8} l_1 l_4(x^2 - 1) + \frac{1}{36} l_1 l_3^2(12x^2 - 7) -$$

$$\left. -\frac{1}{144} l_3 l_4(11x^4 - 42x^2 + 15) + \frac{1}{648} l_3^3(69x^4 - 187x^2 + 52) \right\} +$$

$$+ O(n^{-2}); \tag{2.32}$$

to the order in n considered, x will be normally distributed.

Writing $x - y = h(y + x - y)$, we invert (2.32) to give

$$x - y = \left\{ l_1 + \frac{1}{6} l_3(y^2 - 1) \right\} + \left\{ \frac{1}{2} l_2 y + \frac{1}{24} l_4(y^3 - 3y) - \frac{1}{36} l_3^2(2y^3 - 5y) \right\} +$$

$$+ \left\{ \frac{1}{120} l_5(y^4 - 6y^2 + 3) - \frac{1}{6} l_2 l_3(y^2 - 1) - \frac{1}{24} l_3 l_4(y^4 - 5y^2 + 2) + \right.$$

$$\left. + \frac{1}{324} l_3^3(12y^4 - 53y^2 + 17) \right\} + O(n^{-2}) \tag{2.33}$$

which is a more useful form for many purposes. Terms of order n^{-2} are given in Cornish and Fisher (1937) and Kendall and Stuart (1969, pp. 165–6). The later Fisher and Cornish paper (1960) gives tables of percentage points.

2.9 Estimation

Shenton (1951) investigated the efficiency of the moment estimators for the Gram–Charlier Type A series and found that when four moments are used to fit curves like (2.24), the asymptotic efficiency exceeds 80 per cent only for $\beta_1 < 0\cdot1$, $0 < \beta_2 - 3 < 0\cdot35$ at best. This conclusion is similar to that of Fisher for the Pearson curves, though the efficiency is slightly better for Pearson's Type IV when $3 < \beta_2 \leqslant 3\cdot6$.

As an alternative one might use the first two moments and then use two sample quantiles to obtain values for the coefficients in the series. If $N(>2)$ quantiles are used, we have N equations in the parameters, e.g. the Type A has $p_i = \Phi(x_i) + k_3 H_2(x_i) + k_4 H_3(x_i)$, $i = 1, 2, \ldots, N$, and we select values of k_3, k_4 to minimise some measure of the discrepancy between the theoretical and observed proportions. A formal method based on "least-squares" is often appropriate, but the choice of quantiles is a matter for further investigation.

2.10 An application

The series expansions are often used to approximate sampling distributions, when a good fit in the tails is important for hypothesis testing. The large weight given to the tails by the higher order moments may thus be appropriate for constructing the equation of the approximating curve.

Example 2.1 (Fellingham and Stoker, 1964)

Given a sample of size n, $(x_1, x_2, \ldots, x_n)$ the Wilcoxon test for symmetry about zero uses

$$T = \sum_{i=1}^{n} \{2u(i) - 1\} r_i, \qquad (2.34)$$

where r_i is the rank of x_i, $|x_i|$ being arranged in ascending order, and $u(i) = 1$, $x_i > 0$
$= 0$, $x_i < 0$.

Under the null hypothesis that the distribution is symmetric, the odd order cumulants of T are zero, while

$$\text{var}(T) = \sigma^2 = n(n + 1)(2n + 1)/6 \qquad (2.35)$$

$$\kappa_{2j}(T) = 4^j (4^j - 1) B_{2j} \sum_{i=1}^{n} r_i^{2j}/2j, \qquad (2.36)$$

where B_{2j} are the Bernoulli numbers, e.g. $B_2 = 1/6$, $B_4 = -1/30$, $B_6 = 1/42$. Using the Edgeworth expansion with $z = (T - 1)/\sigma$, the $-1/\sigma$ term being a continuity correction for T, we find the distribution function for z is

$$G(z) = \Phi(z) + c_4 H_3(z) \, \alpha(z) + O(n^{-\frac{1}{2}}) \qquad (2.37)$$

where $$c_4 = \frac{-(3n^2 + 3n - 1)}{10n(n + 1)(2n + 1)}. \qquad (2.38)$$

The usual normal approximation has errors for $n = 100$ comparable to those for $n = 15$ using the Edgeworth/Gram–Charlier form; the latter is accurate enough for practical applications when $n \geqslant 15$, while tables exist for smaller n. The maximum absolute error at $n = 15$ is 25×10^{-5} for z in the range $(1 \cdot 75, 3 \cdot 1)$.

Although κ_4 is always negative here, so that the Edgeworth curve is on the borders of the positive definite region, the results nevertheless suggest the approximation is useful. Fellingham and Stoker also consider the next term in the series, when the maximum error at $n = 15$ falls to 6×10^{-5} (see Exercise 2.3).

2.11 Transformations of the random variable

By expressing the random variable as a series in terms of a normal variate, the Cornish—Fisher series is essentially transforming the original variable to a "near-normal" form; the closeness of the transform depends on the number of terms taken. This approach has several attractions, notably that evaluation of the distribution function is straightforward, since tables of the normal integral and allied functions are readily available.

If a closed functional form, rather than a series, can be used to transform the variable, and tables of this function are readily available, so much the better. The logarithm, scaling down large values of the variable and converting the range from $[0, \infty)$ to $(-\infty, \infty)$ is an obvious candidate.

Early workers in this field were Edgeworth (1898) who considered a series form $x = a + by + c(d + y)^2 + \ldots$, where y is $N(0, 1)$, and Kapteyn (1903), who considered $y = (x + a)^\lambda$, a transformation more recently considered by Box and Cox (1964) for stabilising variances. Haldane (1937) also used this power transformation to achieve approximate normality, selecting a and λ so that β_1 and β_2 were close to the normal values 0 and 3. A normal approximation to the chi-square distribution, constructed by this means, appears in Table 8.1 (section 8.5).

Use of the log-transform came even earlier, however, when Galton (1879) and McAlister (1879) applied the logarithmic transform to data concerned with vital statistics. This distribution, the lognormal, with $\ln y$ normally distributed, has important applications in such diverse fields as economics and astronomy, as witnessed by the monograph (Aitchison and Brown, 1957) devoted to it. In view of the comprehensive discussion to be found there, our treatment is relatively brief.

The best-known example of a transformation using the logarithm of a function of the original variable is probably Fisher's (1921b) form for the correlation coefficient

$$z = \tfrac{1}{2} \ln (1 + r)/(1 - r) = \tanh^{-1}(r). \tag{2.39}$$

2.12 The Johnson system

Johnson (1949a,b) presented a unified discussion of some logarithmic transforms, which allow a wide variety of curve shapes while using only well-tabulated functions.

Writing the density function as

$$\alpha(h) = \delta h_y (2\pi)^{-\frac{1}{2}} \exp \left[-\tfrac{1}{2} \{\delta h(y) + \gamma\}^2 \right], \tag{2.40}$$

where $z = h(y)$ is $N(0, 1)$ and $y = (x - a)/b$, the Johnson transforms are S_L, lognormal:

$$h(y) = \ln y, \quad a < x < \infty \tag{2.41}$$

S_B, bounded range:

$$h(y) = \ln \{y/(1 - y)\}$$
$$= 2 \tanh^{-1} (2y - 1), \quad a < x < a + b \tag{2.42}$$

S_U, unbounded range:

$$h(y) = \ln \{y + (y^2 + 1)^{\frac{1}{2}}\}$$
$$= \sinh^{-1} y, \quad -\infty < x < \infty. \tag{2.43}$$

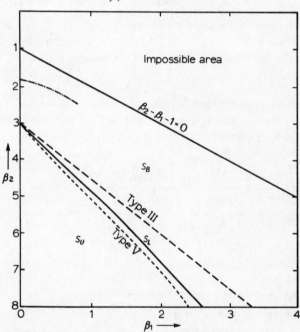

Fig. 2.2   The β_1, β_2 chart for the Johnson system of curves
The lognormal line is marked S_L.
– – – – –   Pearson Type III.
---------------   Pearson Type V.
·—·—·—·—·   Boundary of bimodal curves of system S_B.
(Adapted from Johnson, 1949a)

The choice of S_L and S_B follows from the importance of the lognormal and (2.39); while S_U is more arbitrary, the choice is supported by the fact that these three forms account for the whole β_1, β_2 region – see Fig. 2.2. The curves thus provide as wide a system as the Pearson curves, although the type (lognormal) with only one finite terminal is restricted to a line rather than an area in the β_1, β_2 plane.

The modes of the curves may be found from $(d \ln f)/dy = 0$, excluding the extremes of the range; this gives

S_L: $1 + \gamma\delta + \delta^2 \ln y = 0$; unimodal, mean > mode, +ve skewness;

S_U: $y + \gamma\delta + \delta(y^2 + 1)^{\frac{1}{2}} \ln \{y + (y^2 + 1)^{\frac{1}{2}}\} = 0$; unimodal if $\gamma > 0$,
  mode > median implying –ve skewness, and vice versa;

S_B: the intersection of the curves

$$u_1 = 2y - 1 - \gamma\delta \quad \text{and} \quad u_2 = \delta^2 \ln \{(y/(1 - y)\}; \quad (2.44)$$

the slopes of the curves are 2 and $\delta^2/y(1 - y)$. These will just touch when $2y = 1 + (1 - 2\delta^2)^{\frac{1}{2}}$ and $u_1 = u_2$. This gives us conditions for bimodality which are necessary and sufficient; if

$$\delta < 2^{-\frac{1}{2}}, \quad |\gamma| < \delta^{-1}(1 - 2\delta^2)^{\frac{1}{2}} - 2\delta \tanh^{-1}(1 - 2\delta^2)^{\frac{1}{2}}, \quad (2.45)$$

the curve is bimodal; otherwise the curve is unimodal. The boundary for the bimodal region is shown in Fig. 2.2, and maximum $|\gamma|$ values for given δ are as in Table 2.1, reproduced from Johnson (1949a).

Table 2.1

δ	Maximum $|\gamma|$	δ	Maximum $|\gamma|$
0·7	0·0027	0·3	2·12
0·6	0·175	0·2	4·02
0·5	0·533	0·1	9·37
0·4	1·12		

2.13 Estimation for system S_L

Three main methods of estimation were discussed by Johnson (1949a):

(a) the method of moments,

(b) maximum likelihood, and

(c) the use of quantiles;

to which we add (d) the use of frequency moments. These methods are now discussed for each transform in turn.

From (2.41),

$$z = \gamma + \delta \ln (x - a)/b$$
$$= \gamma - \delta \ln b + \delta \ln (x - a), \qquad (2.46)$$

so that the S_L system has only three identifiable parameters, namely $\delta = \sigma^{-1}$, a, and $\lambda = \gamma - \delta \ln b$.

(a) Moments

Putting $\omega = \exp (1/2\delta^2)$, $\rho = \exp (- \lambda/\delta)$, we find that

$$\left.\begin{aligned}
\mu_1' &= \omega\rho + a \\
\mu_2 &= \omega^2 \rho^2 (\omega^2 - 1) \\
\beta_1 &= (\omega^2 - 1)(\omega^2 + 2)^2 \\
\beta_2 &= \omega^8 + 2\omega^6 + 3\omega^4 - 3.
\end{aligned}\right\} \qquad (2.47)$$

Taking $t = (\omega^2 - 1)^{\frac{1}{2}}$, the expression for β_1 yields

$$t^3 + 3t - \sqrt{\beta_1} = 0, \qquad (2.48)$$

which may be solved using tables provided by Yuan (1933) to give ω, whence

$$\rho = \{\mu_2/t^2(t^2 + 1)\}^{\frac{1}{2}} \qquad (2.49)$$
$$a = \mu_1' - \sqrt{\mu_2}/t, \qquad (2.50)$$

from which a, λ, δ follow. When a is known, only the first two moments are required. These estimators are fully efficient with $\sigma^2 = 0$, but with $\sigma^2 = 0.4$ the efficiency for σ^2 is only 50 per cent (see Fig. 2.3, page 41).

(b) Maximum likelihood

The estimates $\hat{\lambda}, \hat{\sigma}$ follow as the solutions of the equations

$$n\hat{\lambda} = \sum u_i \qquad (2.51)$$
$$n(\hat{\lambda}^2 + \hat{\sigma}^2) = \sum u_i^2, \qquad (2.52)$$

where $u_i = \ln (x_i - a)$, all summations being over $i = 1, \ldots, n$. If a is unknown, we must add the further equation

$$(\hat{\sigma}^2 + \hat{\lambda}) \sum (x_i - \hat{a})^{-1} = -\sum u_i (x_i - \hat{a})^{-1}. \qquad (2.53)$$

Cohen (1951) eliminated $\hat{\lambda}, \hat{\sigma}$ to obtain

$$\sum (x_i - \hat{a})^{-1} \left\{ n \sum u_i (u_i + 1) - \left(\sum u_i \right)^2 + n^2 u_i \right\} = 0. \qquad (2.54)$$

Even for samples of moderate size, (2.54) can be sensitive to changes in a, while the computing time required to solve for a may be considerable. To ease the burden on computing, Cohen (1951) suggests using the smallest order statistic $x_{(1)}$ to find an estimate for a, based on the modified (2.53)

$$x_{(1)} - a_c = \exp(\lambda_c + \sigma_c^2). \tag{2.53'}$$

Further investigation of this approach is required, but Aitchison and Brown (1957) suggest that the resulting estimates are more stable, particularly when using grouped data.

(c) *Quantiles*

By analogy with the normal distribution, if

$$p = \text{prob}(x_p - \lambda = k_p \sigma \leqslant X)$$

$$x_p = \ln y_p = \lambda + k_p \sigma, \tag{2.55}$$

and we may solve for λ, σ. When a is known, Aitchison and Brown found that the pair of percentiles (27, 73) gave 65–81 per cent asymptotic efficiency for λ, while the pair (7, 93) were at least 65 per cent efficient for σ, regardless of the value of σ — see Fig. 2.3.

For a unknown, we could use (2.53') in conjunction with the quantile pairs mentioned above, requiring the solution of the equations

$$e^{\lambda} = \frac{y_{0.73} - y_{0.27}}{\sinh k_{0.73} \sigma} = \frac{y_{0.93} - y_{0.07}}{\sinh k_{0.93} \sigma}, \tag{2.56}$$

whence a can be found from (2.53'). Munro and Wixley (1970) have devised a set of linear estimators for the three-parameter lognormal using order statistics. Their method produces more stable results than (2.54), and the small-sample variances of their estimators compare favourably with those for other methods.

Alternatively, we may use a combination of several methods — Kemsley (1952) used the sample mean and two percentiles. A full discussion of various other alternatives appears in Aitchison and Brown.

(d) *Frequency moments*

Using the theory of section 1.10, we find that the frequency moments for the lognormal are

$$\omega_j = (2\pi b^2)^{(1-j)/2} \, j^{-\frac{1}{2}} \exp\{(1-j)a + (j-1)^2 b^2/2j\}. \tag{2.57}$$

Putting $u_j = \ln \omega_j$ and taking $j = 3/2, 2$, we find that

$$\sigma^2 = 12 \left(u_2 - 2u_{3/2} + \tfrac{1}{2} \ln 1 \cdot 125\right) \tag{2.58}$$
$$(0 \cdot 058\ 882)$$

$$\lambda + \ln \sigma = 2u_2 - 6u_{3/2} - \ln \frac{27}{16} - \frac{1}{2} \ln 2\pi. \qquad (2.59)$$

$$(-1\cdot442\ 187)$$

These estimators have zero efficiency when $\sigma^2 = 0$, but the efficiency increases with σ^2 until $\sigma^2 \sim 6$. The efficiencies of the different methods are plotted in Fig. 2.3.

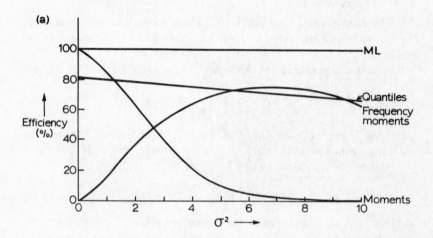

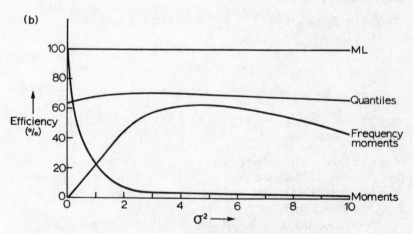

Fig. 2.3 The efficiency of different estimators for the lognormal, lower end-point known: (a) for λ, (b) for σ^2

(Adapted from Fig. 5.4 of Aitchison and Brown, 1957)

Although the asymptotic efficiency of the frequency moments is generally lower than for the quantiles, the difference is not large when $4 \leqslant \sigma^2 \leqslant 8$, and the frequency moments have the advantage of being completely unaffected by location changes — or even knowledge of the location parameter; so we can estimate σ^2, λ, separately and then use these values to estimate a.

2.14 The system S_B

The moments are complicated to evaluate algebraically, and both the moments and the maximum likelihood methods will usually be intractable.

If percentage points are used, we must solve equations of the form

$$k_p = \gamma + \sigma \ln \left(\frac{y - a}{a + b - y} \right), \tag{2.60}$$

following from (2.42) and (2.55). If one or both end-points are known, the solution is straightforward; otherwise we must solve iteratively.

The frequency moments may be found by evaluating

$$\omega_j = (2\pi\sigma^2)^{-j/2} E \left[\{2 \cosh (x/2)\}^{2j-2} \right], \tag{2.61}$$

where x is $N(\mu, \sigma j^{-\frac{1}{2}})$; this is readily solved when $2j$ is an integer. We could then use, say, $\omega_{3/2}$ and ω_2 in conjunction with two quantiles.

2.15 The system S_U

Taking $\omega = \exp(-1/2\delta^2)$, $\rho = -\gamma/\delta$, the moments are

$$\begin{aligned}
\mu_1' &= -\omega \sinh \rho \\
\mu_2 &= \tfrac{1}{2}(\omega^2 - 1)(\omega^2 \cosh 2\rho + 1) \\
\mu_3 &= -\tfrac{1}{4}\omega(\omega^2 - 1)\{\omega^2(\omega^2 + 2)\sinh 3\rho + 3\sinh \rho\} \\
\mu_4 &= \tfrac{1}{8}(\omega^2 - 1)^2 \{\omega^4(\omega^8 + 2\omega^6 + 3\omega^4 - 3)\cosh 4\rho \\
&\quad + 4\omega^4(\omega^2 + 2)\cosh 2\rho + 3(2\omega^2 + 1)\}
\end{aligned} \tag{2.62}$$

from which we could evaluate estimates of the parameters. To facilitate this, Johnson (1949a) gave an abac for ρ, δ in terms of β_1, β_2, and later (Johnson, 1965) produced tables giving γ, δ for $\beta_1 = 0.05\ (0.05)$ 2, β_2 within $3.2\ (0.2)\ 15.0$.

The ML method is again virtually intractable, while the quantiles could be used in the usual way. The frequency moments have the form

$$\omega_j = (2\pi\sigma^2)^{-j/2} E \{(2 \cosh x)^{1-j}\}, \tag{2.63}$$

where x is $N(\mu, \sigma j^{-\frac{1}{2}})$, which can only be evaluated, for j an integer, in series form. However, it can be shown that if

$$C = \frac{27\,\omega_3\,\omega_{3/2}^4}{16\sqrt{3}\,\omega_2^4} \qquad (2.64)$$

then $C = 1$ if the curve is lognormal or normal,

> 1 if the curve is S_B,

< 1 if the curve is S_U,

enabling a choice to be made between the different curves.

2.16 The Burr system

Our discussion so far has covered only those systems which have the normal curve at their centre. The important sampling properties of the normal justify this emphasis, but other approaches are possible.

Rather than use the density function, Burr (1942) used a differential equation involving the distribution function, i.e.

$$\frac{dF}{dx} = A(F)\,g(x). \qquad (2.65)$$

We may then choose an appropriate form of $A(F)$, giving the basic curve of the system when $g(x) \equiv 1$. For example, we could choose

(a) $A(F) = F(1 - F)$, giving the logistic curve with

$$F(x) = \left\{1 + \exp\left(\frac{a - x}{b}\right)\right\}^{-1}, \quad -\infty < x < \infty. \qquad (2.66)$$

Such a system would have applications in bioassay.

(b) $A(F) = 1 - F$, when $F(x) = 1 - e^{-x}$, $\qquad (2.67)$

with applications in life-testing (see Exercise 2.7). Burr used (2.66), giving the general equation

$$F(x) = [1 + \exp\{-G(x)\}]^{-1}, \qquad (2.68)$$

where $G(x) = \int_{-\infty}^{x} g(x)\,dx$, and selected the form

$$F(x) = 1 - (1 + x^a)^{-b}, \quad 0 \leqslant x < \infty, \qquad (2.69)$$

which has been further studied by Hatke (1949) and Burr and Cislak (1968); see also Exercise 4.11. Burr and Cislak give the range of possible (β_1, β_2) values for distribution (2.69), and their results are presented in modified form in Fig. 2.4. From their small-sample studies

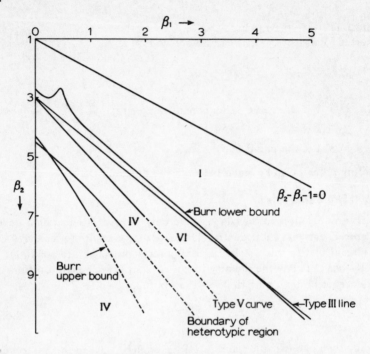

Fig. 2.4   The β_1, β_2 chart for the Burr curves
(Adapted from Burr and Cislak, 1968)

of the mean and the median, it would appear that the mean is usually a better estimator for the location parameter. Burr (1968) gives the small-sample properties of the range.

Another convenient form based on (2.66), giving a finite range curve, is

$$g(x) = h(x - a)^{-1} + k(b - x)^{-1}, \qquad (2.70)$$

yielding

$$F/(1 - F) = (x - a)^h (b - x)^{-k}, \quad a \leqslant x \leqslant b. \qquad (2.71)$$

When $h = k = 1$ we have the uniform, while for $h = k = 3$ we get the cube law (Kendall and Stuart, 1950).

Efficient estimators for the parameters of (2.69) and (2.71) are not available, but sample moments and quantiles seem to offer the most practicable approach. Hassanein (1969) has derived best linear unbiased estimators for the parameters of the logistic distribution, based on up to nine sample quantiles.

2.17 Perks' distributions

The density function

$$h(x) = \sum_{r=0}^{m} a_r \, e^{-rx} \Big/ \sum_{r=0}^{n} b_r \, e^{-rx} \qquad (2.72)$$

was proposed by Perks (1947). In applications, the particular form

$$h(x) = ae^{-x}/(b_0 + b_1 e^{-x} + b_2 e^{-2x})$$

is sufficiently general. Talacko (1958) considered the symmetric form

$$h(x) = a/(k + 2\cosh x), \quad k > -2, \qquad (2.73)$$

which contains the logistic ($k = 2$). The system has been used in actuarial work.

2.18 Examples of fitted curves

To conclude our survey of the systems of univariate continuous curves we give the results of fitting curves from three different systems to data on the length of beans (Table 2.2). The Gram–Charlier and

Table 2.2
Distribution of length of beans and various fitted curves

Length of beans (mm) class midpoints	Observed frequency	Gram–Charlier A	Pearson Type IV	Johnson S_U
> 9·25 ⎫	—	⎧	1·9	2·6
9·5 ⎬	1	⎱ 0·9	2·6	2·7
10·0 ⎭	7	⎩	5·4	5·8
10·5	18	5·9	11·3	12·1
11·0	36	29·6	24·2	25·7
11·5	70	98·7	52·5	55·2
12·0	115	206·2	113·8	118·0
12·5	199	258·7	243·7	249·3
13·0	437	280·7	503·6	508·7
13·5	929	713·4	968·9	970·6
14·0	1787	1788·4	1638·9	1642·5
14·5	2294	2593·0	2229·8	2240·6
15·0	2082	2155·4	2132·6	2130·3
15·5	1129	1012·7	1181·6	1151·5
16·0	275	241·7	299·3	290·1
16·5 ⎫	55	25·6	28·5	32·2
17·0 ⎬	6	12·8	⎧ 1·4	2·0
> 17·25 ⎭	—	16·3	⎩	0·1
Total	9440	9440·0	9440·0	9440·0
Value of χ^2	—	344	103	87

(Reproduced by permission from Pretorius, 1930 and Johnson, 1949a)

Pearson Type IV were fitted by Pretorius (1930) and the S_U curve by Johnson (1949a). The χ^2 values are reported to guide comparison only.

The fit was by moments in all cases, and the moments yield $\beta_1 = 0\cdot829\ 136$, $\beta_2 = 4\cdot862\ 944$. Referring to Fig.2.1, we see that this is just inside the positive definite region for the Gram–Charlier Type A (2.24) but outside for the Edgeworth form (2.25).

EXERCISES

2.1   When only the first observed moment is equated to the theoretical moment in the Edgeworth expansion, show that the coefficients become

$$a_0 = 1, \quad a_1 = 0, \quad a_2 = (\kappa_2 - \eta_2)/2!,$$
$$a_3 = \kappa_3/3!, \quad a_4 = \{\kappa_4 + 3(\kappa_2 - \eta_2)^2\}/4!,$$
$$a_5 = \{\kappa_5 + 10\kappa_3(\kappa_2 - \eta_2)\}/5!,$$
$$a_6 = \{\kappa_6 + 10\kappa_3^2 + 30\kappa_4(\kappa_2 - \eta_2) + 15(\kappa_2 - \eta_2)^3\}/6!.$$

2.2   If the density function $g(x)$ can be approximated by the series

$$a(x)\left\{1 + \sum_{j=3}^{k} c_j H_j(x)/j!\right\},$$

$a(x)$ being the standardised normal density function, show that the distribution function is

$$G(x) = \Phi(x) - a(x)\sum_{j=2}^{k-1} c_{j+1} H_j(x)/j!$$

where $\Phi(x) = \int_{-\infty}^{x} a(y)\,dy.$

2.3   Extend the results of section 2.10 to show that the next terms in the series for T are

$$-\frac{4(3n^4 + 6n^3 - 3n + 1)}{35\{n(n+1)(2n+1)\}^2} H_5(x)\,a(x) - \frac{1}{200}\left\{\frac{3n^2 + 3n - 1}{n(n+1)(2n+1)}\right\}^2 H_6(x)\,a(x).$$

(Fellingham and Stoker, 1964)

2.4   Show that the logarithms of the frequency moments, $u_j = \ln \omega_j$, of the lognormal in section 2.12 have asymptotic variance

$$\text{var}(u_j) = \frac{j^2}{n}\left[j(2j-1)^{-\frac{1}{2}} \exp\left\{\frac{(j-1)^2\sigma^2}{j(2j-1)}\right\} - 1\right]$$

and covariance

$$\text{cov}\,(u_j, u_k) = \frac{jk}{n}\left[\left(\frac{jk}{j+k-1}\right)^{\frac{1}{2}} \exp\left\{\frac{(j+k)(j-1)(k-1)\sigma^2}{2jk\,(j+k-1)}\right\} - 1\right].$$

Hence derive the variances of the estimators for λ, σ in (2.58) and (2.59).

2.5 Using the first two terms of the Edgeworth (or Gram–Charlier A) series, establish the approximate relationship

$$3\,(\text{mean} - \text{median}) \simeq \text{mean} - \text{mode}.$$

(Haldane, 1942)

2.6 Using the Burr distribution (2.69) with $a = 4\cdot874$, $k = 6\cdot158$, show that x has the moments $\mu_1' = 0\cdot644\,693$, $\mu_2 = 161\cdot984$, $\beta_1 = 0\cdot000$, $\beta_2 = 3\cdot000$, so that this form may be used as an approximation to the normal distribution function. (Burr, 1967)

2.7 The family $dF/dx = (1 - F)\,g(x)$, defined in section 2.16, may be described as a family of failure-time distributions (x being the time to failure of a component). $g(x)$ is then the hazard function or failure rate. Show that the following distributions are members of this class, and consider their plausibility as failure-time distributions:

exponential, $\quad\quad\quad\quad\quad\quad e^{-x}, \quad x \geqslant 0$

Weibull, $\quad\quad\quad\quad\quad\quad\quad bx^{b-1} \exp\,(-x^b), \quad x \geqslant 0$

truncated extreme value, $\quad \exp\,(x + 1 - e^x), \quad x \geqslant 1$

Pareto, $\quad\quad\quad\quad\quad\quad\quad bx^{-b-1}, \quad x \geqslant 1.$

Finally, find the distribution with hazard function $1 - (1 + x)^{-1}$.

2.8 The class of symmetric stable distributions has characteristic functions of the form

$$\phi(t) = -|t|^a, \quad 0 < a \leqslant 2.$$

When $a > 1$, show that the density function is of the form

$$f(x) = \frac{1}{\pi a} \sum_{j=0}^{\infty} (-1)^j \frac{\Gamma\{(2j+1)/a\}}{(2j)!} x^{2j}.$$

(Bergström, 1952; see also Fama and Roll, 1968)

CHAPTER 3

MULTIVARIATE SYSTEMS OF CURVES

3.1 Introduction

Once we allow that univariate data may come from a non-normal population we must accept that the multivariate normal model is likewise inadequate; Pearson (1895) realised that if data followed a skew curve, the classical Galton–Dickson theory of correlation would need to be modified.

For convenience we deal mainly with the bivariate problem, but the extension will usually be obvious (or clearly impractical). Two questions must be answered:

(a) what is the form of the multivariate density function?

(b) what is the regression and correlation structure of this function?

Two important properties of bivariate normal regressions are linearity of the regression function and homoscedasticity (constant variance); it is useful to know whether either of these properties holds for more general systems.

We will discuss various bivariate generalisations of systems discussed earlier and see how the above questions are answered. Pretorius (1930) reviewed work in the field up to that date, giving numerous examples of fitting data and the resulting regression curves.

For a more detailed discussion of the distributions considered in this chapter, see the monograph by Mardia (1970).

3.2 The Pearson system

Having derived his univariate differential equation from the hypergeometric distribution, Pearson obtained a pair of equations in a similar way from the double hypergeometric series, yielding for the density function $f(x, y)$

$$\left. \begin{array}{l} \dfrac{1}{f} \dfrac{\partial f}{\partial x} = \dfrac{\text{cubic in } x, y}{\text{quartic in } x, y} \\[2ex] \dfrac{1}{f} \dfrac{\partial f}{\partial y} = \dfrac{\text{another cubic in } x, y}{\text{same quartic in } x, y} \end{array} \right\} \qquad (3.1)$$

subject to $(\partial^2 f / \partial x \partial y) = (\partial^2 f / \partial y \partial x)$.

Without further limitations on the constants, progress was impossible. In imposing restrictions, it seemed reasonable to restrict the regression function in some way — to be linear or quadratic, or to have a constant (or linear) scedastic function, for example.

Many of these conditions impose relationships between higher order moments. For example, Pearson (1923) discussed the bivariate beta curve known as the Dirichlet distribution, which has

$$f(x, y) \; \alpha \; (x/b_1)^{p_1} \; (y/b_2)^{p_2} \; (1 - x/b_1 - y/b_2)^q \qquad (3.2)$$

with correlation

$$\rho(x, y) = -\left\{ \frac{(p_1 + 1)(p_2 + 1)}{(p_1 + q + 2)(p_2 + q + 2)} \right\}^{\frac{1}{2}} \qquad (3.3)$$

but subject to the restriction

$$\frac{\beta_{20} + 3}{\beta_{10} + 4} = \frac{\beta_{02} + 3}{\beta_{01} + 4},$$

where $\beta_{10} = \mu_{30}^2/\mu_{20}^3$, etc., μ_{ij} being $E\{(x - \mu_x)^i \; (y - \mu_y)^j\}$.

The marginal curves of (3.2) are Type I curves, while the regression and scedastic functions are linear. Other properties of this distribution are given by Mauldon (1959).

Isserlis (1914) fitted a double hypergeometric series to data concerning two hands at whist, and Wicksell (1917) showed the Isserlis series to have linear regressions (in the sense that the points fall upon a straight line); Pearson (1923) showed that the scedastic function was quadratic, but also that the general equations (3.1) lead to a cubic regression function. The extent of linearity for regression functions is clearly of interest, and we now present a theorem of Steyn which allows us to evaluate the regression function directly from the characteristic function.

Consider a k-variate distribution $x_1, \ldots, x_k$ with regression function for x_1 given $x_2, \ldots, x_k$

$$\tilde{x}_1 = E(x_1 \mid x_2, \ldots, x_k) \equiv E(x_1 \mid \mathbf{x}^*); \qquad (3.4)$$

the density function is $dP(\mathbf{x})$ with characteristic function $\phi \equiv \phi(\theta_1, \ldots, \theta_k)$ while the conditional density is $dP(x_1 \mid \mathbf{x}^*)$. The Stieltjes form of the integral is used and s_i denotes the region of integration for x_i.

Theorem 3.1 (Steyn, 1957)

We may represent the regression function by

$$\tilde{x}_1 = \frac{Q(x_2, \ldots, x_k)}{R(x_2, \ldots, x_k)} = \frac{Q}{R} \qquad (3.5)$$

where Q and R are given by

$$\left[\frac{\partial}{\partial \theta_1} \{R(D_2, \ldots, D_k)\phi\}\right]_{\theta_1 = 0} = [Q(D_2, \ldots, D_k)\phi]_{\theta_1 = 0}, \qquad (3.6)$$

$D_i = (\partial/\partial\theta_i)$; the converse holds.

Proof (3.5) $\Rightarrow$ (3.6)

From (3.4),

$$E(x_1 \mid \mathbf{x}^*) = \int_{s_1} x_1 \, dP(x_1 \mid \mathbf{x}^*);$$

if (3.5) holds, then

$$\int_{s_1} x_1 \, R \, dP(x_1 \mid \mathbf{x}^*) = \int_{s_1} Q \, dP(x_1 \mid \mathbf{x}^*).$$

Multiplying by $\exp(\theta_2 x_2 + \ldots + \theta_k x_k) \, dP(\mathbf{x}^*)$, where $dP(\mathbf{x}^*)$ is the marginal density, and integrating over $\{s_2, \ldots, s_k\} \equiv \mathbf{s}$,

$$\int_{\mathbf{s}} \ldots \int e^{(\boldsymbol{\theta}^*)' \mathbf{x}^*} x_1 \, R \, dP(\mathbf{x}) = \int_{\mathbf{s}} \ldots \int e^{(\boldsymbol{\theta}^*)' \mathbf{x}^*} Q \, dP(\mathbf{x}). \qquad (3.7)$$

Now (Widder, 1946, p. 63), the Laplace transform has the property that if

$$\int \ldots \int e^{\boldsymbol{\theta}' \mathbf{x}} \, dP(\mathbf{x}) = \phi(\boldsymbol{\theta})$$

then $\qquad \int \ldots \int e^{\boldsymbol{\theta}' \mathbf{x}} g(\mathbf{x}) \, dP(\mathbf{x}) = \{g(D_1, \ldots, D_k)\, \phi(\boldsymbol{\theta})\}, \qquad (3.8)$

from which (3.6) follows.

(3.6) $\Rightarrow$ (3.5) Conversely, the uniqueness of the Laplace transform means that we may invert (3.8) to obtain $g(\mathbf{x}) \, dP(\mathbf{x})$. Applying this result to (3.6) shows that (3.5) holds almost everywhere.

The importance of this result is that for any density function we may set up a differential equation for the characteristic function and use this to derive the regression equation — this is applied to discrete distributions in section 7.4.

Example 3.1 (Steyn, 1960)

Consider the multivariate beta (Dirichlet)

$$f(\mathbf{x}) \propto (x_1/b_1)^{p_1} \ldots (x_k/b_k)^{p_k} (1 - \sum_i x_i/b_i)^q. \qquad (3.9)$$

Differentiating with respect to x_1, we obtain

$$x_1(1 - \sum_i x_i/b_i)\frac{\partial f}{\partial x_1} = p_1(1 - \sum_i x_i/b_i)f - qx_1 f/b_1,$$

the transform of which is (see Exercise 3.1)

$$- D_1(\theta_1\phi) + \sum_{i=1}^{k} \bar{b}_j^1 D_1 D_j(\theta_1\phi)$$
$$= p_1\phi - p_1 \sum_{i=1}^{\bar{k}} \bar{b}_j^1 D_j\phi - \bar{b}_1^1 q_1 D_1\phi.$$

Rearranging terms and putting $\theta_1 = 0$,

$$[(q + 2)D_1\phi]_{\theta_1=0} = b_1[(p_1 + 1)\{1 + \sum_{j=2}^{k} \bar{b}_j^1 D_j\}\phi]_{\theta_1=0}. \qquad (3.10)$$

Applying Theorem 3.1 to (3.10),

$$(q + 2)\tilde{x}_1 = b_1(p_1 + 1)(1 - \sum_{j=2}^{k} x_j/b_j), \qquad (3.11)$$

generalising Wicksell's (1917) result.

Example 3.2 (Steyn, 1960)

Suppose the density function satisfies

$$\frac{1}{f}\frac{\partial f}{\partial x_i} = \frac{a_{i0} + \mathbf{a}_i'\mathbf{x}}{b_{i00} + \mathbf{b}_{i0}'\mathbf{x} + \mathbf{x}'\mathbf{B}_i\mathbf{x}}, \qquad (3.12)$$
$$i = 1, 2, \ldots, k,$$

where $\mathbf{a}_i$, $\mathbf{b}_{i0}$ are $(k \times 1)$ vectors, and $\mathbf{B}_i$ are $(k \times k)$ symmetric matrices with elements b_{ilj}.

Applying the theorem as before, we find that

$$(a_{ii} + 2b_{iii})\tilde{x}_i = -(a_{i0} + b_{ii0}) - \sum_{j\neq i} (a_{ij} + 2b_{ilj})x_j, \qquad (3.13)$$

showing that this generalisation of the Pearson system leads to linear regression functions. The theorem could likewise be applied to establish Pearson's result for system (3.1). The theorem may be readily extended to cover $E(\tilde{x}_1^2)$, so that the variance function may be derived.

3.3 Other Pearson system approaches

Continuing the research of Pearson's Statistical Laboratory team into bivariate systems, Rhodes (1922) obtained the form

$$f(x, y) \propto e^{-lx-my}(1 - x/a + y/b)^p(1 + x/c - y/d)^q \qquad (3.14)$$

which satisfies (3.12), so that the regression is linear; other properties are discussed in the original paper. While a variety of special cases were considered in the literature, a general approach was lacking until Narumi (1923); (he also showed that only the bivariate normal had straight-line homoscedastic regression functions for both variables).

By imposing various conditions on the regression functions and working back to the density functions within the system (3.1) Narumi obtained two main classes, those with (a) linear regression and linear heteroscedasticity — essentially a special case of Example 3.2 above, and (b) non-linear homoscedastic regression. The general form for this case is

$$f(x, y) \; \alpha \; \exp \{-\gamma (X - a)(Y - b) + c_1 x + c_2 y\}, \qquad (3.15)$$

where $X = e^{l_1 x}$, $Y = e^{l_2 y}$; the regression functions $\tilde{X}$, $\tilde{Y}$ are given by

$$\frac{A_1 l_2}{\tilde{X}} = b - Y, \quad \frac{A_2 l_1}{\tilde{Y}} = a - X, \qquad (3.16)$$

where A_i depends on γ, c_1, c_2. The marginal and conditional curves for X, Y are Pearson's Type III.

3.4 Series expansions

Starting from the bivariate normal density with means zero, variances one, correlation ρ, i.e. $\alpha(x, y \mid \rho) \equiv \alpha(x, y)$, we could derive a series expansion for a general bivariate density $g(x, y)$ with standardised means and variances, as

$$g(x, y) \; = \; \alpha(x, y) + \sum\sum_{i+j \geqslant 3} (-1)^{i+j} \frac{a_{ij}}{i! \; j!} D_x^i \, D_y^j \, \alpha(x, y), \quad (3.17)$$

where $D_x = \partial/\partial x$, $D_y = \partial/\partial y$. Using the marginal densities $\alpha_i(.)$, $i = 1, 2$, rather than $\alpha(x, y)$, we obtain

$$g(x, y) \; = \; \alpha_1(x)\alpha_2(y) + b_{11} D_x D_y \alpha_1 \alpha_2 + \sum\sum_{i+j \geqslant 3} (-1)^{i+j} \frac{b_{ij} D_x^i \, \alpha_1 \, D_y^j \, \alpha_2}{i! \; j!}.$$

$$(3.18)$$

This form is more convenient in applications as it involves only the univariate Hermite polynomials, which are tabulated — see Appendix B.

If the random variables are not standardised by mean and standard deviation, the summations in (3.17), (3.18) could be extended to cover $i + j \geqslant 1$.

Suppose the (i, j)th cumulant κ_{ij} is $O(n^{1-(i+j)/2})$, n a parameter, then the coefficients for the Gram–Charlier Series A form of (3.17) to $O(n^{-1})$ are

$$a_{ij} = \kappa_{ij}, \quad i + j = 3, 4. \tag{3.19}$$

The Edgeworth form of (3.17) given by Kendall (1949) has the additional terms, to $O(n^{-1})$,

$$\left. \begin{array}{ll} l_{60} = \frac{1}{2} l_{30}^2 & l_{51} = l_{30} l_{21} \\[2mm] l_{42} = l_{30} l_{12} + \frac{1}{2} l_{21}^2 \\[2mm] l_{33} = l_{30} l_{03} + l_{21} l_{12} \end{array} \right\} \tag{3.20}$$

where $l_{ij} = a_{ij}/i!\,j!$, other terms following by symmetry.

The form (3.18) has $b_{ij} = a_{ij}$, except that $b_{11} = \kappa_{11} = \rho$, $b_{22} = \kappa_{22} + 2\kappa_{11}^2$.

Rhodes (1925) discussed the properties of (3.17) when $i + j = 3$, while Pearson (1925) derived the coefficients for his 15-parameter surface using the terms $x^i y^j$ rather than the Hermite polynomials.

The regression function for the Gram–Charlier form of (3.17), for y given x, is

$$\tilde{y} = \rho x + \frac{(\kappa_{21} - \rho\kappa_{30}) H_2(x)/2 + (\kappa_{31} - \rho\kappa_{40}) H_3(x)/720}{1 + \kappa_{30} H_3(x)/6 + \kappa_{40} H_4(x)/24}. \tag{3.21}$$

This, the scedastic curve, and similar expressions for the Edgeworth form were found by Wicksell (1917). If (3.18) is used, the regression function may be derived using the properties of Hermite polynomials: let the marginal density of x be

$$\{1 + \sum_{j=3}^{k} c_{j0} H_j(x)/j!\}\, a(x) \tag{3.22}$$

– the form (2.24) or (2.25), for example; then

$$\tilde{y} = \frac{\sum_{j=1}^{k-1} b_{j1} H_j(x)/j!}{1 + \sum_{j=3}^{k} c_{j0} H_j(x)/j!}. \tag{3.23}$$

In particular, the Gram–Charlier A form is, for terms with $i + j \leqslant 4$,

$$\tilde{y} = \frac{\kappa_{11} x + \kappa_{21} H_2(x)/2 + \kappa_{31} H_3(x)/6}{1 + \kappa_{30} H_3(x)/6 + \kappa_{40} H_4(x)/24}. \tag{3.24}$$

Taking κ_{i+j} as $O(n^{1-(i+j)/2})$, we see that, to terms of $O(n^{-\frac{1}{2}})$, (3.21) is a quadratic in x, but (3.24) contains terms up to x^4. Unlike the Pearson system, where a significant subclass of the bivariate forms yields linear regressions, none of these forms do, except as special cases.

Although the bivariate normal has been used throughout this section, other distributions may be used, such as the gamma (Krishnamoorthy and Parthasarathy, 1951; Moran, 1969). Expansion (3.18) is more useful in such cases.

3.5 Existence of series

Chambers (1967) gives extensions of the results in section 2.3 for the validity of multivariate Edgeworth series. Taking u to be the standardised means of n identically distributed X_i with characteristic function $\phi_X(\theta)$, we can say that if $a(x)$ is the standardised k-variate normal density, and $g_n(x)$ is an arbitrary density function, n a parameter, say sample size, then —

(1) provided the moment vectors $\mu_1, \mu_2, \ldots, \mu_{r+2}$ exist with

$$\int |\phi_X(\theta)|^\nu \, d\theta < \infty \quad \text{for some } \nu \geqslant 1, \tag{3.25}$$

then for $n \geqslant \nu$, and as $n \to \infty$,

$$\epsilon(x) = g_n(x) - \{1 + \sum_{j=3}^{r} n^{1-k/2} \Pi(-D_m)^{im}\} a(x) \tag{3.26}$$

is $o(n^{-r/2})$. Note that $D_m = \partial/\partial x_m$, and that Π is taken over $i_1 + \ldots + i_k = j$.

(2) provided the first $(r + 3)$ moment vectors exist and

$$\int_{\text{all } \theta > kn} a |\phi_X(\theta)| \, d\theta = o(n^{\delta - (r+1)/2}) \tag{3.27}$$

$$k > 0, \quad 0 < a < \tfrac{1}{2},$$

the error in (3.26) is $o(n^{\delta - (r+1)/2})$.

(3) provided the first $(r + 2)$ moments exist and

$$\lim \sup |\phi_X(\theta)| < 1, \text{ for some } \nu \geqslant 1, \tag{3.28}$$

the error in (3.26) is $O\{(\ln n)^{(k+1)/2} n^{-(r+1)/2}\}$.

For further discussion, see Chambers (1967) and Feller (1966, p. 506).

3.6 Bivariate transformation systems

If we transform two variables x, y so that $s = s(x)$, $t = t(y)$ are normally distributed, it is clear that the joint distribution of s, t need not be the usual form of the bivariate normal. However, we now follow

Johnson (1949b) and assume that the usual bivariate form holds, then proceeding to investigate the regression equations of the Johnson system.

Using the Johnson transforms of section 2.12, the regression function of y on x would be given by

$$\bar{y} = \{2\pi(1 - \rho^2)\}^{-\frac{1}{2}} \int_{-\infty}^{\infty} y(t) \exp \{-(t - \rho s)^2/2(1 - \rho^2)\} \, dt. \quad (3.29)$$

This will usually be in too complicated a form to be useful. Johnson therefore suggested using the median regression, i.e. if $s - \lambda_1 = \sigma_1 g_1(x)$, with t similarly defined, the median regression of y on x is $t = \rho s$,

i.e.
$$g_2(y^*) = \frac{\rho \sigma_1}{\sigma_2} g_1(x) + (\rho \lambda_1 - \lambda_2)/\sigma_2; \quad (3.30)$$

this will be linear when $\rho \lambda_1 = \lambda_2$, $\rho \sigma_1 = \sigma_2$ and $g_1 \equiv g_2$, or when x, y are both normally distributed. This is analogous to the result for "mean regression" — that both regression lines are linear for non-perfectly correlated variables only when both x and y are normal.

Example 3.3

When x is s_N, $g_1(x) = x$ and y is s_B, $g_2(y) = \ln\{y/(1 - y)\}$, the median regression function is the logistic form

$$y^* = (1 + ae^{-bx})^{-1} \quad (3.31)$$

where $a = \exp\{(\lambda_2 - \rho \lambda_1)/\sigma_2\}$

$b = \rho \sigma_1/\sigma_2$.

Some other forms are

$$\left. \begin{array}{l} s_N \text{ on } s_L: y^* = \ln a + b \ln x, \\ s_L \text{ on } s_L: y^* = -ax^b, \text{ etc.} \end{array} \right\} \quad (3.32)$$

A full list is given by Johnson (1949b, p. 299). These forms are used in economics — the lognormal curve is a useful representation of the distribution of personal incomes (Aitchison and Brown, 1957), and both x and y are usually random variables in such cases.

The drawback of this approach is that the form of the regression function is determined by the marginal curves rather than the bivariate data. The advantages of the transformation method are that
  (1) tables of the bivariate normal distribution and of the "transla-
      tion" functions are all that is required for fitting purposes;

(2) the number of parameters (four for each marginal curve, one for the correlation between the transformed variates) is not too unwieldy.

3.7 Canonical variables

Given a bivariate distribution $G(x, y)$ with marginals $F_1(x)$, $F_2(y)$, Van Uven (1925) developed a graphical method of analysing correlated variables. If x, y are not normally distributed, we construct two functions $s(x)$, $t(y)$ of x and y which follow the normal law and give as high a measure of correlation as possible. This approach was formalised by Lancaster (1958) who defined sets of canonical variables $\{s_i(x)\}$, $\{t_i(y)\}$ such that

$$\left.\begin{array}{c} s_i \equiv s_i(x), \quad t_i \equiv t_i(y) \quad s_0 \equiv t_0 \equiv 1 \\[4pt] \iint s_i \, s_j \, dG(x, y) = \iint t_i \, t_j \, dG(x, y) = 0, \quad i \neq j \\ = 1, \quad i = j \end{array}\right\} \quad (3.33)$$

and
$$\iint s_i \, t_i \, dG(x, y) = \rho_i \quad \text{for } i, j = 1, 2, \ldots$$

These variables can be made to obey the further relation

$$\iint s_i \, t_j \, dG(x, y) = 0, \quad i \neq j, \qquad (3.34)$$

suggesting the representation

$$dG(x, y) = \left\{1 + \sum_{i=1}^{r} \rho_i \, s_i \, t_i\right\} dF_1(x) \, dF_2(y) \qquad (3.35)$$

with "mean square contingency" $\phi_i^2 = \sum_{i=1}^{r} \rho_i^2$.

This enables us to construct distributions with arbitrary marginal and correlation structures, and also provides an alternative goodness-of-fit test for the usual form of the bivariate normal. The regression structure of these models has been explored by Eagleson and Lancaster (1967).

3.8 Bivariate systems with general margins

In addition to the canonical correlation approach of the previous section, various systems have been proposed to cover bivariate curves with general marginal distributions, instead of basing the development on the normal. Let $F_1(x)$, $F_2(y)$ denote the marginal distributions of $G(x, y)$.

(a) Frechet (1951) showed that if

$$G_0(x, y) = \max \{F_1(x) + F_2(y) - 1, 0\}$$

and $$G_1(x, y) = \min \{F_1(x), F_2(y)\}$$

then G_0 and G_1 are valid bivariate distributions, and any other bivariate distribution G, with margins F_1 and F_2, satisfies

$$G_0(x, y) \leqslant G(x, y) \leqslant G_1(x, y).$$

Fréchet considered

$$G = \lambda G_0 + (1 - \lambda) G_1, \quad 0 \leqslant \lambda \leqslant 1, \tag{3.36}$$

but this form suffers from the weakness that $G \neq F_1 F_2$ when the variates are independent.

(b) Morgenstern (1956),

$$G(x, y) = F_1(x) F_2(y) [1 + \alpha \{1 - F_1(x)\} \{1 - F_2(y)\}], \tag{3.37}$$

yielding the density

$$g(x, y) = f_1 f_2 \{1 + \alpha (1 - 2F_1)(1 - 2F_2)\}, \tag{3.38}$$

clearly requiring $|\alpha| \leqslant 1$.

Farlie (1960) discusses the more general form

$$G(x, y) = F_1(x) F_2(y) \{1 + \alpha A_1(x) A_2(y)\} \tag{3.39}$$

considered in section 3.9.

(c) Gumbel (1960a, 1961),

$$\{- \ln G(x, y)\}^m = \{- \ln F_1(x)\}^m + \{- \ln F_2(y)\}^m \tag{3.40}$$

with the limiting form, as $m \to \infty$,

$$G(x, y) = \min \{F_1(x), F_2(y)\}$$

corresponding to (3.36) with $\lambda = 0$.

This is followed up by Gumbel and Mustafi (1967), who discuss bivariate extremal distributions in detail. Gumbel also considers (3.37) as a bivariate form for the logistic curve; for the direct generalisation of the logistic, see Exercises 3.4 and 3.5.

(d) Plackett (1965), Mardia (1967). If

$$\psi = \frac{G(1 - F_1 - F_2 + G)}{(F_1 - G)(F_2 - G)}, \quad 0 \leqslant \psi < \infty, \tag{3.41}$$

we may use the positive root for G given F_1, F_2 and ψ. Mardia showed that one, and only one, root is always permissible. When $F_1 = F_2 = \frac{1}{2}$,

$G = \psi^{\frac{1}{2}}/2(1 + \psi)^{\frac{1}{2}}$, while more generally

$$G = F_1 F_2, \quad \psi = 1$$
$$= [s - \{s^2 - 4\psi(\psi - 1)F_1 F_2\}^{\frac{1}{2}}]/2(\psi - 1), \quad \psi \neq 1 \quad (3.42)$$

where $s = 1 + (\psi - 1)(F_1 + F_2)$.

3.9 The Farlie–Gumbel system

Daniels (1944) proposed a general class of measures of correlation and disarray, which includes the product-moment, Spearman's rho and Kendall's tau as special cases. For the relation between x_i and y_j assign the scores a_{ij}, b_{ij} respectively, then the coefficient is of the form

$$\frac{\Sigma a_{ij} \, b_{ij}}{\Sigma a_{ij}^2 \, \Sigma b_{ij}^2}.$$

Considering this coefficient in relation to the bivariate system (3.39), Farlie (1960) showed the product moment coefficient to be an efficient estimator for α in (3.39) when

$$A_i(v) = \int_{-\infty}^{v} u dF_i(u) / \int_{-\infty}^{v} dF_i(u) \quad \text{for } i = 1, \; v = x \text{ and } i = 2, \; v = y.$$

$$(3.43)$$

Similarly, the rank coefficients are efficient when

$$A_i(v) = 1 - F_i(v), \quad (3.44)$$

i.e. for the special case (3.37) which includes Gumbel's bivariate logistic. However, the condition $|\alpha| \leqslant 1$ implies that $|\rho| \leqslant 3/\pi^2$, a severe restriction on the correlation. In this case, the conditional density for y given x is

$$g_2(y \mid x) = \frac{e^{-y}}{(1 + e^{-y})^2}\left(1 + \alpha \frac{(1 - e^{-x})(1 - e^{-y})}{(1 + e^{-x})(1 + e^{-y})}\right) \quad (3.45)$$

with moment generating function

$$M(\theta) = \pi\theta(1 + \beta\theta)/\sin \pi\theta, \quad (3.46)$$

where $\beta = \alpha(1 - e^{-x})(1 + e^{-x})^{-1}$.

The regression function is therefore

$$\tilde{y} = \beta = 2\alpha(1 - e^{-x})^{-1} - \alpha, \quad (3.47)$$

providing an alternative model for the "logistic" regression curve (cf. section 3.6, equation 3.31). The variance function is

$$\text{var}(y \mid x) = \pi^2/3 - \beta^2 \quad \text{(cf. Exercises 3.4 and 3.5)}.$$

If the random variables are transformed so as to be logistic in form, the "mean regression" functions cannot be found in general, but the median regression is, from (3.45),

$$y* = \ln \{(1 + \beta^2)^{\frac{1}{2}} - \beta\} = -\sinh^{-1}\beta, \tag{3.48}$$

from which median regression functions may be developed for transformed variables as in section 3.6.

3.10 The Plackett system

The review given below is very brief; for a more detailed discussion see Mardia (1970, especially pp. 55–73). Mardia (1967) showed that the Plackett system is the only one which can both (a) satisfy the Frechet bounds given in section 3.8 (as $\psi \to 0$ and infinity respectively), and (b) yield $G = F_1 F_2$ when the random variables are independent (at $\psi = 1$).

From (3.42), the non-central moments are given by

$$\mu'_{rs} - \mu'_r \mu'_s = \int_{-\infty}^{\infty} x^r \int_{-\infty}^{\infty} y^{s-1} \frac{\partial}{\partial x} (G - F_1 F_2) \, dy \, dx, \tag{3.49}$$

provided that (Mardia, 1967)

(i) $\lim_{x \to \pm\infty} x^r (G - F_1 F_2) = 0$

(ii) $\lim_{y \to \pm\infty} y^s \{G_2(y \mid x) - F_2(y)\} = 0$

(iii) $\int_{-\infty}^{\infty} |y|^{s-1} \cdot |G_2(y \mid x) - F_2(y)| \, dy < I(x)$, where $I(x)$ is an integrable function of x; and the conditional distribution of y given x is

$$G_2(y \mid x) = \frac{F_2 \psi - G(\psi - 1)}{1 + (F_1 + F_2 - 2G)(\psi - 1)}. \tag{3.50}$$

When F_1, F_2 are normal, the resulting bivariate normal, $c(x, y; \psi)$, is not the usual form $N(x, y; \rho)$, although the two agree closely when

$$\rho = \cos \{\pi/(1 + \psi^{\frac{1}{2}})\}; \tag{3.51}$$

for further discussion of this point, see Mardia (1967).

In the general case, for fitting ψ, Plackett proposed

$$\hat{\psi}_1 = (ad)/(bc), \tag{3.52}$$

where a, b, c, d are the frequencies of pairs (x_i, y_i) in the quadrants $(x \leqslant p, y \leqslant q)$, $(x \leqslant p, y > q)$, $(x > p, y \leqslant q)$ and $(x > p, y > q)$, that is, by using an estimator of the tetrachoric correlation. Mardia shows that

$\hat{\psi}_1$ is most efficient when p, q are chosen to coincide with the population medians, $F_1(p) = F_2(q) = \frac{1}{2}$. However, he goes on to show that
the sample correlation between F_1 and F_2 has an asymptotic efficiency
between 1·7 and 2·2 times that of (3.52) for ψ, and he gives tables to
aid the computation.

Using (3.42), a pair of correlated random variables may be constructed for particular marginals — this is important for simulation
studies and is a useful property of the system.

3.11 The multivariate exponential family

By analogy with section 1.12, the density function of the multivariate exponential (ME) family is of the form

$$f(x, \theta) = \exp\{x'\,\theta + A(x) + B(\theta)\}, \tag{3.53}$$

after transformations of θ and x, if necessary. Provided the range of x
is independent of θ, the maximum likelihood estimators are the solutions to

$$\bar{x}_j = B_j(\theta); \quad j = 1, \ldots, k, \tag{3.54}$$

where $B_j(\theta) = (\partial/\partial\theta_j)B(\theta)$ and $\bar{x}_j$ is the sample mean for the jth
variate.

Several properties of this family of distributions are now listed.

Property ME 1 (cf. Property E 1). The characteristic function of $f(x, \theta)$
is

$$\phi(t) = \exp\{-B(\theta) + B(\theta + i\,t)\}. \tag{3.55}$$

Property ME 2 (Patil, 1965a) (cf. Property E 2). A necessary and
sufficient condition for a distribution to be ME is that its cumulants
obey the relationships

$$\kappa(r_1, r_2, \ldots, r_i + 1, \ldots, r_k) = \frac{\partial\kappa(r_1, \ldots, r_k)}{\partial\theta_i},$$

where $\kappa(i, j, \ldots, m) = D_1^i D_2^j \ldots D_k^m\{B(\theta + i\,t)\}$, and $D_i = \partial/\partial t_i$.

Property ME 3 (cf. Property E 5). Given the ML equations

$$\frac{\partial \ln L(\theta)}{\partial\theta_j} = \bar{x}_j - B_j(\theta); \quad j = 1, \ldots, k,$$

it can be seen that a function $\tau(\theta)$ can be minimum variance unbiased
(MVU) estimable if and only if

$$\tau(\theta) = a + \sum_{j=1}^{k} b_j B_j(\theta).$$

For further discussion on MVU estimators, see section 7.6 on the power series distributions.

3.12 Fitting bivariate distributions and regression functions

We have said very little about general fitting problems so far, for several reasons. First, the maximum likelihood and similar methods are generally intractable, even with modern high-speed computing equipment. Secondly, where transformed data are used, a reasonable procedure will often be to fit the marginals separately and then to estimate the "correlation" parameter(s). However, adequate fits for the margins do *not* ensure that the bivariate model will be a reasonable representation of the population. Finally, for the series expansions, the sample moments (or k-statistics) offer the only feasible approach in most cases.

More important than curve-fitting would seem to be fitting the regression functions. For the Pearson system, when the regression is linear, the obvious method would be to use least squares, but, as is well known, these estimators may be biased when the regressors are random variables. Nevertheless, in view of the computational difficulties and low efficiency of moment estimators such a method has attractions.

When transformations are used, one is tempted to apply the maximum likelihood method to the transformed data. This again may lead to peculiar results if the transforms have been selected using the data; but the absence of any alternative suggests that this approach will be used, with an appeal to the panacea of "large samples" until better procedures are available.

It is important to remember that regression functions, as defined in this chapter, are median or expected value functions, $E(y \mid x)$, and the basis of these models (all variables random) is different from that of the least-squares model (independent variables known). The two methods are therefore complementary, and in some fields where the least-squares assumptions cannot be realistically applied, the expected value models might repay further study.

3.13 Multivariate extensions

Where possible, extensions have been outlined in the appropriate sections of the chapter. For regression purposes, Steyn's theorem is a useful tool, particularly for the Pearson system; the median regression curves of Johnson can also be extended.

62

EXERCISES

3.1 If the Laplace transform of f is

$$L(f) \equiv \phi = \int e^{\theta x} f(x) \, dx,$$

show that $L(f') = -\theta \phi$ and $L(xf) = \phi'$.

3.2 Using Steyn's theorem (section 3.2), find the regression equation for

$$f(x, y) \propto (1 + b_1 x^2 + b_2 y^2 + 2b_3 xy)^{-n}.$$

(Pearson, 1923)

3.3 Establish the form of the median regression functions for the bivariate Johnson system in section 3.6. (Johnson, 1949b)

3.4 Taking the bivariate logistic distribution as

$$G(x, y) = (1 + e^{-x} + e^{-y} + be^{-x-y})^{-1},$$

show that the joint density function is

$$g(x, y) = G^3 e^{-x-y} \{(1 + be^{-x})(1 + be^{-y}) + 1 - b\}.$$

Hence find the conditional density $g_2(y \mid x)$ and show that its moment generating function is

$$\frac{\pi\theta}{\sin \pi\theta} c^{\theta-1} \{b + (1 - b)(1 + \theta)(1 + e^{-x})^{-1}\},$$

where $c = (1 + be^{-x})(1 + e^{-x})^{-1}$.

3.5 Using the results of Exercise 3.4, show that the regression function of y on x is

$$\tilde{y} = (1 - b)u - \ln \{1 + (b - 1)u/b\},$$

where $u = (1 + be^{-x})^{-1}$.

Show also that

$$\text{var}(\tilde{y}) = \frac{\pi^2}{3} - (1 - b)^2 u^2.$$

3.6 For the exponential distribution, $F(x) = 1 - e^{-x}$, use the bivariate form (3.37) to show that the correlation coefficient in this case cannot exceed an absolute value of $0 \cdot 25$. (Gumbel, 1960a)

3.7 Find the density function $g(x, y; \psi)$ for the Plackett system, from (3.42). Define the random variable Z by $F_1(Z) = 1 - F_1(X)$ and let $g^*(z, y; \psi)$ denote the joint density function of Z and Y. Show that

$$g^*(z, y; \psi) = g(z, y; 1/\psi).$$

Further, when both f_1 and f_2 are symmetric, show that the moments satisfy

$$\mu_{rs}(\psi) = \begin{cases} \mu_{rs}(1/\psi), & r \text{ and } s \text{ even} \\ -\mu_{rs}(1/\psi), & r \text{ and } s \text{ odd} \\ 0, & \text{otherwise.} \end{cases}$$

(Mardia, 1967; Steck, 1968)

CHAPTER 4

MIXTURES OF DISTRIBUTIONS

4.1 Introduction

If $\mathcal{F} = \{F_\lambda.; \lambda \in \Lambda\}$ is a family of distributions, each member of the family being indexed by the parameter λ, we may define a mixture for $\mathcal{F}$ with respect to some distribution $G(\lambda)$, or with respect to G-measure, by

$$H(x) \equiv H_G(x) = \int_\Lambda F(x; \lambda) \, dG(\lambda), \qquad (4.1)$$

where $G \in \mathcal{G}$, the set of *non-degenerate* probability distributions $\{G(\lambda)\}$, and λ may be s-dimensional, $\lambda = \{\lambda^1, \lambda^2, \ldots, \lambda^s\}$ defined over Λ. The integral (4.1) is defined in the Stieltjes sense.

We are, in effect, "averaging" $F(x; \lambda)$ over the values of $\lambda \in \Lambda$. $H(x)$ is called a *mixture*, or more specifically a G-mixture, while $G(\lambda)$ is the *mixing distribution*. The general theory of mixtures was developed by Robbins (1948), Robbins and Pitman (1949), and Teicher (1954a), although the basic idea goes back to Karl Pearson (1894) — see section 4.8.

Intuitively we think of a mixture as two or more populations physically mixed (such as the gases in the atmosphere), where both the population distribution and the proportion of the total formed by any population are unknown. However, this is only one aspect of the topic, and there are several types of mixture, namely

(a) *Finite mixtures*

There are k distinct populations occurring in proportions $p_i (p_1 + p_2 + \ldots + p_k = 1)$, each with distribution function $F_i(x)$. The distribution function for the mixture, $H(x)$, is

$$H(x) = p_1 F_1(x) + \ldots + p_k F_k(x), \qquad (4.2)$$

the $F_i(x)$ relating to the *components* of the mixture.

(b) *Countable mixtures*

Let k be countable rather than finite in (4.2), i.e.

$$H(x) = p_1 F_1(x) + p_2 F_2(x) + \ldots, \qquad (4.3)$$

then (4.3) is a countable mixture.

The most important examples of this type occur when the p_i form a standard discrete distribution, such as the Poisson, i.e. a Poisson mixture. We might regard $H(x)$ as the distribution for sampling in two stages — selecting first-stage units with probability p_i, which contain second-stage units following a distribution $F_i(x)$; however, the p_i are usually unknown.

(c) If the points of increase of $G(\lambda)$ are non-countable then $G(\lambda)$ will usually be absolutely continuous, except perhaps at one or two points (e.g. zero flows in dam theory).

The present definition of a mixture is a broad one and certain special cases have attracted considerable attention:

Definition 4.1

If
$$H(x) = \int_T F(x - y) \, dG(y), \quad y \in T, \tag{4.4}$$

we call $H(x)$ the *convolution* of F and G. This may be written
$$F(x - y) * G(y).$$

Definition 4.2

The G-mixture
$$H(x) = \int_T F(x \mid cy) \, dG(y), \quad y \in T, \tag{4.5}$$

c a constant, is also referred to as a *compound distribution*; written $X \wedge Y$ in the notation of Gurland (1957). Y is the *compounder*, and $G(y)$ the *compounding distribution*. The class of distribution $\mathcal{H} = \{H(x)\}$, defined by the distribution functions $F \in \mathcal{F}$, is referred to as the class of $\mathcal{F}$-compound distributions.

In conjunction with this we also have

Definition 4.3

Let Y_1, Y_2 have probability generating functions $P_1(z)$, $P_2(z)$, respectively. Then the random variable $X = Y_1 \vee Y_2$ has probability function
$$P(z) = P_1[P_2(z)] \tag{4.6}$$

and X follows a *generalised distribution*. Y_2 is the *generaliser*, and may be continuous or, more usually, discrete.

This terminology follows Feller (1943 — definition 4.2), Gurland (1957), and Katti and Gurland (1961), but it is by no means universally accepted. Feller (1957) refers to (4.6) as a compound density, while Kendall and Buckland (1960) hold the description of (4.5) as compound to be unsatisfactory; they also prefer Feller's 1957 usage. This problem of appropriate definitions was recently aired by Goodhardt (1966) and Adelson (1966a, b).

The generalised distributions, as defined by (4.6), are, when $P_1(z)$ is discrete, known as *contagious* distributions or *true contagion* models (see Section 6.5). By contrast, the compound distributions with $F(x)$ discrete are *apparent contagion*, or *accident proneness*, models (see section 6.6).

The main need is to avoid further confusion over these terms, which seems best achieved by accepting the majority opinion as embodied in definitions 4.2 and 4.3, although the terms "G-mixture" and "true contagion model" will often suffice. After listing a few properties of mixtures, we consider the identifiability of parameters for estimation purposes. The remainder of this chapter is devoted to estimation problems for finite mixtures, the discussion of other (discrete) models being left until Chapter 6.

4.2 Properties of mixtures

A few properties of mixtures are listed below, drawn from Teicher (1960):

Property M1. When $G(\lambda)$ is absolutely continuous the density function $h(x)$ is given by $\int f(x; \lambda) \, dG(\lambda)$.

Property M2. If $F(x; \lambda)$ is absolutely continuous then, by Fubini's theorem, so is $H(x)$.

Definition 4.4

A family of distributions is said to be *additively closed* if the convolution

$$F(x; \lambda) * F(x; \mu) = F(x; \lambda + \mu) \tag{4.7}$$

or, for characteristic function $\phi(t; .)$, if

$$\phi(t; \lambda) \, \phi(t; \mu) = \phi(t; \lambda + \mu). \tag{4.8}$$

Property M3. If H_i is a G_i mixture of an additively closed family $\mathcal{F}$, $i = 1, 2$, then $H_1 * H_2$ is a $G_1 * G_2$ mixture of $\mathcal{F}$. The proof follows from consideration of the characteristic function of $H_1 * H_2$.

Property M4 (partial converse of M3). If, for $r \geq 1$, G_1 and G_2 have exactly r points of positive mass and the convolution of a G_1 mixture of $\mathcal{F}$ with a G_2 mixture of $\mathcal{F}$ is a $G_1 * G_2$ mixture of $\mathcal{F}$, then $\mathcal{F}$ is additively closed.

A corollary of M3 is that the convolution of two compound Poisson distributions is itself a compound Poisson distribution with mixing distribution the convolution of the two original mixing distributions. M3 permits the immediate evaluation of the sampling distribution of the sum of n observations.

A class of distributions of considerable theoretical interest is that of infinitely divisible distributions (I.D.). A distribution is said to be I.D. if its characteristic function, $\phi(t)$, is such that

$$\psi(t) = \{\phi(t)\}^{1/k}$$

is a characteristic function for all $k > 0$. The canonical representation of I.D. distributions is now given.

Definition 4.5

The function $\phi(t)$ is the characteristic function of an I.D. distribution $F(x)$ with mean μ and variance σ^2 if and only if

$$\ln \phi(t) = i\mu t + \int_{-\infty}^{\infty} (e^{itx} - 1 - itx)\frac{1}{x^2} dK(x),$$

where $K(x)$ is a non-decreasing function and $K(\infty) - K(-\infty) = \sigma^2$.

This representation was first given by Kolmogorov (1932). For a full discussion of I.D. distributions, see Robinson (1959), Chapter 3. It is readily established that the normal and the Poisson are infinitely divisible distributions. The relevance of this to our present discussion lies in

Property M5 (Maceda, 1948). In a mixture, if the mixing distribution is infinitely divisible, then so is the mixture.

Thus, whenever the mixing distribution is I.D. we can immediately derive the sampling distribution for the sum of n observations from the mixture.

Property M6. The mixture density $h(x)$ may be written

$$h(x) = \int_{\Lambda} f(x; \lambda) \, dG(\lambda),$$

whence the posterior density $dK(\lambda \mid x)$ is

$$dK(\lambda \mid x) = \frac{f(x; \lambda) \, dG(\lambda)}{h(x)},$$

providing an obvious link between Bayes' theorem and mixtures.

4.3 Identifiability of mixtures

Before we can discuss estimation or hypothesis testing for a mixture we must know whether the mixture is uniquely determined; that is, whether the parameters are *identified*. Failing this, estimation may be impossible. However, in some cases, lack of identifiability may be overcome. See, for example, the estimators for finite mixtures of exponential-type distributions in section 4.5 *et seq.*

The remainder of the present section follows Teicher (1961) and the review by Blischke (1965).

Definition 4.6

The mixture $H \equiv H_G$ is said to be *identifiable* if the relationship

$$H(x) = \int F(x; \lambda) \, dG(\lambda)$$
$$= \int F(x; \lambda) \, dG^*(\lambda) \qquad (4.9)$$

implies that $G = G^*$ for all $G^* \in \mathcal{G} \cup \mathcal{I}$, where $\mathcal{I}$ is the class of distributions with unit measure assigned to a single point.

By suitably restricting $\mathcal{F}$ or $\mathcal{G}$ we can ensure identifiability, but this is clearly undesirable unless the restrictions are mild.

The class $\mathcal{H} = \{H(x)\}$ is identifiable if every member of the class is. A consequence of this is that no identifiable G-mixture, defined for $G \in \mathcal{G}$, of $\mathcal{F} = \{F(\lambda), \lambda \in \Lambda\}$ can again be a member of $\mathcal{F}$.

We now review the main results on the identifiability of mixtures, indicating their implications. Distribution functions are taken to be continuous on the right, i.e. $F(x) = \text{prob} \, (X \leqslant x)$.

Theorem 4.1 (Teicher, 1961)

The mixture H in (4.1) is identifiable if λ is scalar and
either $F(x; \lambda)$ is additively closed, while λ ranges over (a) non-negative integer, (b) non-negative rational, or (c) non-negative real, numbers;

or λ is a translation parameter, unless the Fourier transform of $F(x; \lambda)$ is identically zero in some non-degenerate real interval;

or $F(0; \lambda) = 0$, unless the Fourier transform of $F(e^x; \lambda)$ is identically zero in some non-degenerate real interval.

Example 4.1

We apply the theorem to mixtures involving the binomial as mixing distribution:

$$\text{prob}\,(X \leqslant x) \equiv F(x; n, p) = \sum_{j \leqslant x} \binom{n}{j} p^j q^{n-j}. \qquad (4.10)$$

(a) When p is fixed, $F(x; n)$ is additively closed, since

$$\phi(t; n) = (q + pe^{it})^n, \qquad (4.11)$$

so that $\phi(t; n_1)\, \phi(t; n_2) = \phi(t; n_1 + n_2)$, where n ranges over the positive integers.

(b) However, when n is fixed, none of the above conditions are satisfied and the class of $G(p)$ mixtures is not identifiable.

The class of compound Poisson distributions is, however, identifiable (see Exercise 4.1), so that this class does not contain the Poisson itself; the class of normal mixtures with mixing density $dG(\sigma^2)$ for the scale parameter is also identifiable (see section 4.12).

4.4 Mixtures of binomial distributions

Example 4.1 shows that members of this class are not always identifiable; it is therefore of interest to ask under what conditions identifiability is achieved.

Let

$$f(r; n, p) = \binom{n}{r} p^r q^{n-r}, \quad r = 0, 1, \dots, n, \qquad (4.12)$$

the usual binomial density. Then, for the mixing distribution g_n, $n = r$, $r + 1, \dots$, the resulting compound density is

$$h(r) = \sum_{n=r}^{\infty} g_n f(r; n, p) \qquad (4.13)$$

Define the generating function

$$G(z; \theta) = \sum_n g_n(\theta) z^n; \qquad (4.14)$$

then the mixture density, $h(r)$, has generating function $G(q + pz; \theta)$. Rao (1965) interpreted $\{g_n\}$ as the distribution of numbers initially observed in an investigation, while $f(r; n, p)$ yields the number of (undamaged) observations finally recorded.

Example 4.2

When $g_n = \theta^n e^{-\theta}/n!$, i.e. n is Poisson,

$$G(q + pz, \theta) = e^{\theta p (z-1)}$$

and θ, p are confounded, or unidentifiable. The condition for θ and p to be unidentifiable is that

$$G(q + pz; \theta) = \bar{G}(z; \alpha) = \sum_n \bar{g}_n(\alpha) z^n, \qquad (4.15)$$

where $\alpha \equiv \alpha(\theta, p)$. For the special case where $\{g_n\}$ is a power series distribution (see section 6.2), the probabilities are

$$g_n(\theta) = a_n \theta^n/m(\theta); \quad \bar{g}_n(\alpha) = b_n \alpha^n/\bar{m}(\alpha),$$

$\{a_n\}$, $\{b_n\}$ constants; so that condition (4.15) becomes

$$\frac{m(q\theta + pz\theta)}{m(\theta)} = \frac{\bar{m}(\alpha z)}{\bar{m}(\alpha)}. \qquad (4.16)$$

Differentiating successively with respect to z and substituting $z = 0$, we obtain, for the sth derivative,

$$\frac{m^{(s)}(\theta q)}{m(\theta)} \theta^s q^s = \frac{\bar{m}^{(s)}(0)}{\bar{m}(\alpha)} \alpha^s, \quad s = 0, 1, 2, \ldots, \qquad (4.17)$$

yielding the differential equation for θ

$$\frac{m^{(2)}(\theta)}{m(\theta)} = c \left[\frac{m^{(1)}(\theta)}{m(\theta)}\right]^2, \quad c \text{ a constant}. \qquad (4.18)$$

This simplifies to

$$\frac{d \ln m(\theta)}{d\theta} = \frac{1}{c\theta + d}, \qquad (4.19)$$

where c and d are constants. This equation is satisfied by the Poisson, binomial (see Example 4.1) and negative binomial distributions, when the latter has the form (Sprott, 1965)

$$f(r; k, p) = \binom{k + r - 1}{k - 1} \theta^k (1 - \theta)^r.$$

Thus the parameters are unidentifiable in these cases. Sprott (1965) extends Rao's work to consider the general case $G[w(p, z); \theta]$ which is unidentified if there exists an $\alpha \equiv \alpha(\theta, p)$ such that $G[w(p, z); \theta(\alpha, p)]$ is a function of α and z only, not of p. A necessary condition is for the original distribution to have a generating function of the form

$$G(z;\ \theta)\ =\ G[u(\theta)/v(z)]\ . \tag{4.20}$$

Throughout this discussion we have assumed θ, α and p to be scalars, but replacing them by vectors of appropriate magnitude does not alter the conclusion.

4.5 Mixtures for the exponential family

The multivariate form of the exponential family, defined in section 1.12, is

$$f(\mathbf{x}\,|\,\boldsymbol{\theta})\ =\ \exp\left[\ \sum_{j=1}^{m}\ \theta_j c_j(\mathbf{x}) + A(\mathbf{x}) + B(\boldsymbol{\theta})\right], \tag{4.21}$$

where $\boldsymbol{\theta}$ is an $(m \times 1)$ vector of parameters and
$\mathbf{x}$ is an $(n \times 1)$ vector of variables.

If we employ the mixing distribution $G(\boldsymbol{\theta})$, the resulting mixture has density

$$h(\mathbf{x})\ =\ \int f(\mathbf{x}\,|\,\boldsymbol{\theta})\ dG(\boldsymbol{\theta})\,; \tag{4.22}$$

under conditions similar to those of Theorem 4.1, Barndorff-Nielsen (1965) has shown this class of mixtures to be identifiable.

We now restrict our attention to finite mixtures of single parameter exponential-type distributions. Let the class of distributions be

$$\mathcal{H}\ =\ \{H(x);\, H(x)\ =\ \sum_{i=1}^{k}\ p_i F(x,\ \theta_i)\}, \tag{4.23}$$

where p_i are all positive and $\Sigma\, p_i\ =\ 1$. $F(x,\ \theta_i)$ is the distribution function derived from the univariate form of (4.21). The general class of finite one-parameter mixtures has been shown to be identifiable if and only if the $F(x,\ \theta_i)$ provide a basis for k-dimensional space (Yakowitz and Spragins, 1968). This is always so when F is of exponential type.

One estimation problem of considerable interest in empirical Bayesian analysis arises when the $F(x,\ \theta_i)$ are completely known, but the mixing distribution is not. Boes (1966) obtained the minimum variance unbiased estimator for $k = 2$, while methods for any k have been proposed by Choi (1968) and Deely and Kruse (1968). Other possibilities are reported in Rolph (1968) and Yakowitz (1968).

We now turn to the more general problem where both the $\{\theta_i\}$ and the $\{p_i\}$ are unknown. For $k = 2$ several authors have considered particular distributions, and these are listed in Table 4.1. Estimators using the sample moments are given in section 4.6, while a more general approach, for any k, is given in section 4.7. Blischke (1965) provides a review of recent developments in this area.

Table 4.1

Exponential-type distributions: one-parameter finite mixtures

Distribution	Unknown parameters	Reference
Exponential, Gamma, Weibull	Scale	Rider* (1961a,b)
Binomial	Proportion	Rider* (1961a); Blischke* (1964)
Poisson	Mean	Rider* (1961a); Schilling* (1947)
Geometric	Proportion	Daniels (1961)
Normal	(i) Scale, means known and equal	cf. gamma;
	(ii) Mean, variances known and equal	Pearson (1894)

* These authors discuss estimation problems; Blischke also considers maximum likelihood estimators.

4.6   Estimation for mixtures of two populations

Writing the finite mixture in the form

$$h(x \mid \theta_1, \theta_2, \alpha) = \alpha f(x \mid \theta_1) + (1 - \alpha) f(x \mid \theta_2), \qquad (4.24)$$

for the distributions listed in Table 4.1, and for the negative binomial, the method of moments yields

$$\alpha \theta_1^i + (1 - \alpha) \theta_2^i = l_i; \quad i = 1, 2, 3, \qquad (4.25)$$

where l_i is a known (in any particular case) linear function of the ith crude moment (or factorial moment for discrete distributions).

Given these three equations in three unknowns, we may solve for the parameters in terms of the l_i to obtain the estimators (Rider, 1961a)

$$\left. \begin{aligned} \alpha^* &= \frac{l_1 - \theta_2^*}{\theta_1^* - \theta_2^*} \\ v^* &= l_1 u^* - l_2 \\ u^* &= (l_3 - l_2 l_1)/(l_2 - l_1^2), \end{aligned} \right\} \qquad (4.26)$$

where $u^* = \theta_1^* + \theta_2^*, v^* = \theta_1^* \theta_2^*.$ $\qquad (4.27)$

There are certain conditions which the parameters should satisfy, e.g. $0 \leqslant \alpha \leqslant 1$ in all cases, and $\theta_i \geqslant 0$ in all except normal case (ii). However, Rider (1961b) has shown that (a) the estimators given by

(4.26) may fail to satisfy these constraints, and (b) the roots θ_1^*, θ_2^* may fail to exist; both events occurring with positive probability for any sample size.

To overcome these difficulties we may use a linear programming formulation. From the first equation of (4.25) we obtain

$$\alpha(1 - \alpha) = \frac{(l_1 - \theta_2)(\theta_1 - l_1)}{(\theta_1 - \theta_2)^2}, \tag{4.28}$$

which must satisfy $\frac{1}{4} \geqslant \alpha(1 - \alpha) \geqslant 0$. Provided θ_1, θ_2 exist, this double inequality is satisfied when

$$l_1 u - v \geqslant l_1^2. \tag{4.29}$$

Note that in developing the LP formulation we drop all asterisks for convenience. θ_1 and θ_2 exist if

$$u^2 - 4v \geqslant 0;$$

from (4.29), $\quad u^2 - 4v \geqslant u^2 - 4l_1(u - l_1) = (u - l_1)^2 \geqslant 0, \tag{4.30}$

so that (4.29) ensures both the existence of θ_1, θ_2 and admissible α values.

With the exception of normal case (ii) we also require θ_1, θ_2 positive, so we add

$$u \geqslant 0, \quad v \geqslant 0. \tag{4.31}$$

The final formulation becomes

minimise $\quad \sum_{i=1}^{4} c_i d_i$

subject to $\quad \left.\begin{aligned} l_1 u - \quad v + d_1 - d_2 \qquad\qquad &= l_2 \\ l_2 u - l_1 v \qquad\qquad + d_3 - d_4 &= l_3 \\ l_1 u - \quad v \qquad\qquad\qquad &\geqslant l_1^2 \\ \text{all variables} \geqslant 0 \end{aligned}\right\} \tag{4.32}$

When the θ_i refer to proportions, i.e. $0 \leqslant \theta_i \leqslant 1$, we add the further conditions

$$\left.\begin{aligned} 1 - v &\geqslant 0 \\ 1 + v - u &\geqslant 0 \end{aligned}\right\} \tag{4.33}$$

The c_i values in the objective function are at the disposal of the researcher, but it would seem reasonable to take $c_1 = c_2 > c_3 = c_4 > 0$ to give the second crude moment, which has lower variance, higher "weight", and to allow deviations either way equal importance. We may put $c_4 = 1$ without loss of generality. The choice of c_i does not appear to be crucial.

From the basic properties of the LP formulation it is apparent that only one of (d_1, d_2) and only one of (d_3, d_4) will appear in the final solution; while if solution (4.26) yields estimates in the required ranges, they will be the same as those obtained from (4.32). Thus the LP approach is essentially a way of handling the "awkward" cases. It can reduce the variance of the estimators considerably, especially for small samples (see Exercise 4.5).

The solution given by the LP approach will, when it differs from the moments solution, put one of the parameters at the end of its range (e.g. α, $\theta_i = 0$ or 1), but this is less difficult to interpret than the negative estimates which may arise otherwise.

4.7 Estimation for mixtures of several populations

If we allow $k > 2$ in (4.23) a programming approach may still be used, but it cannot be reduced to a set of linear constraints. A general method, using both frequencies and moments, has been developed by Kabir (1968). We repeat the exponential-type density function for convenience:

$$f(x, \theta_i) = \exp\{x\theta_i + A(x) + B(\theta_i)\}.$$

Let the range of x be $(a, b]$, and divide this into $2k$ equal parts at $a = a_0, a_1 = a_0 + c, \ldots, a_{2k} = a_0 + 2kc = b$. If the range is infinite it should be truncated, and if finite, truncation may prove desirable. If the variate is discrete, the range should be split so that each interval contains an equal number of points of increase. In the remainder of the section we consider only x continuous, but the development for x discrete is broadly similar.

Let $$T_j = \int_{a_j}^{a_{j+1}} \exp\{-A(x)\} h(x)\, dx, \quad j = 0, \ldots, 2k - 1, \quad (4.34)$$

then $$T_j = \sum_{i=1}^{k} Q_i \lambda_i^j, \quad \text{where } \lambda_i = \exp\{\theta_i c\} \quad \text{and}$$

$$Q_i = \frac{p_i(\lambda_i - 1)}{\theta_i} \exp\{B(\theta_i) + a_0 \theta_i\}.$$

If $$\prod_{i=1}^{k} (\lambda - \lambda_i) = 0 = \lambda^k - \gamma_1 \lambda^{k-1} \ldots - \gamma_k, \quad (4.35)$$

we see that $$\gamma_k T_0 + \gamma_{k-1} T_1 + \ldots + \gamma_1 T_{k-1} = T_k, \quad (4.36)$$

after some manipulation. Similar equations can be derived for the set

$\{T_i, \ldots, T_{k+i}, i = 0, 1, \ldots, k-1\}$; that is,

$$\sum_{j=1}^{k} \gamma_j T_{k+i-j} = T_{k+i}. \tag{4.37}$$

Provided that the Q_i are non-zero it can be shown that the k equations of this form are linearly independent, so that

$$\gamma = \mathbf{Z}^{-1}\mathbf{T}, \tag{4.38}$$

where $\gamma' = (\gamma_k, \ldots, \gamma_1)$, $\mathbf{Z} = \{z_{ij} = T_{i+j-2}\}$, and $\mathbf{T}' = \{T_k, \ldots, T_{2k-1}\}$. Further, it is easily seen that the jth cumulant of $h(x)$ is

$$\kappa_j = \sum_{i=1}^{k} p_i \frac{\partial^j}{\partial \theta_i^j} \{B(\theta_i)\}, \quad j = 1, \ldots, k-1. \tag{4.39}$$

The first $(k-1)$ cumulants, together with (4.39), provide $(2k-1)$ estimating equations. The κ_j can be consistently estimated by k statistics, while Kabir shows that for the sample set $S = \{x_1, \ldots, x_n\}$, the estimators

$$t_j = \frac{1}{n} \Sigma \exp\{A(x)\}, \quad x \in \{I_j \cap S\}, \tag{4.40}$$

where $I_j = (a_j, a_{j+1}]$, are consistent for T_j (x continuous or discrete). It is worth noting that the θ_i are estimated independently of the p_i. As no constraints are imposed on the estimators, the estimates may lie outside the proper ranges of the parameters.

Example 4.3   (Kabir, 1968)

Consider a finite mixture of two exponential distributions, so that

$$f(x, \theta_i) = \theta_i \exp(x\,\theta_i), \quad i = 1, 2,$$

whence $A(x) = 0$ and $B(\theta_i) = \ln \theta_i$. We take $\theta_1 < \theta_2$. $k = 2$, so we divide the range into four intervals I_0, I_1, I_2 and I_3 say, covering the range $(0, b]$ and containing n observations. From (4.40) it is apparent that

$$t_j = n_j/n,$$

the proportion of observations falling in the jth interval. From (4.37), therefore, we obtain the estimators

$$\begin{pmatrix} \hat{\gamma}_1 \\ \hat{\gamma}_2 \end{pmatrix} = \begin{pmatrix} t_0 & t_1 \\ t_1 & t_2 \end{pmatrix}^{-1} \begin{pmatrix} t_2 \\ t_3 \end{pmatrix}.$$

From (4.35), when $k = 2$, the values of λ are

$$\lambda_1, \lambda_2 = \tfrac{1}{2}\gamma_1 \pm \tfrac{1}{2}(\gamma_1^2 + 4\gamma_2)^{\frac{1}{2}},$$

from which the $\hat{\lambda}_i$ are obtained by substituting $\hat{\gamma}_i$. The $\hat{\theta}_i$ follow directly. Finally, since the first k statistic is the sample mean, $\bar{x}$, we have

$$\hat{p}_1 = \hat{\theta}_1(\bar{x}\hat{\theta}_2 - 1)/(\hat{\theta}_2 - \hat{\theta}_1).$$

A more efficient alternative is to use the ML estimators. An iterative scheme for the evaluation of these estimates is now given, due to Hasselblad (1969). The log likelihood function, for a sample of size n, is

$$L = \sum_{i=1}^{n} \ln h(x_i),$$

where $\quad h(x_i) = \sum_{j=1}^{k} p_j f_j(x_i) \quad$ and $\quad f_j(x_i) \equiv f(x_i, \theta_j).$

Differentiating, we obtain

$$\frac{\partial L}{\partial \theta_j} = \Sigma \, p_j \frac{f_j(x_i)}{h(x_i)} (x_i + b_j), \quad j = 1, \ldots, k; \qquad (4.41)$$

and $\qquad \dfrac{\partial L}{\partial p_j} = \Sigma \, \{f_j(x_i) - f_k(x_i)\}/h(x_i), \quad j = 1, \ldots, k - 1; \qquad (4.42)$

where $b_j = \partial B(\theta_j)/\partial \theta_j$ and all summations are taken over $i = 1, \ldots, n$. Putting the derivatives equal to zero, we obtain

$$b_j = \Sigma \, \{x_i f_j(x_i)/h(x_i)\}/\Sigma \, \{f_j(x_i)/h(x_i)\} \qquad (4.43)$$

and $\qquad np_j = p_j \, \Sigma \, \{f_k(x_i)/h(x_i)\}. \qquad (4.44)$

Equations (4.43) and (4.44) may be solved iteratively, possibly using the moment estimators as starting values, and then substituting the values obtained at the mth stage of the calculation into the right-hand sides to give the values for the $(m + 1)$th stage. Asymptotic variances can then be computed in the usual way.

4.8 Other finite mixtures

When we turn to multi-parameter and/or non-exponential type populations, estimation is more difficult and general methods have yet to be devised. Of particular interest is the finite mixture of the two normal populations, and we now consider this in detail.

Study of the finite mixture of two normal populations, P_1 and P_2, was initiated by Pearson (1894). We write the density function as

$$h(x) = \alpha f_1(x) + (1 - \alpha) f_2(x), \qquad (4.45)$$

where, for normal population P_j,

$$f_j(x) = (2\pi\sigma_j^2)^{-\frac{1}{2}} \exp\left[-(x - \theta_j)^2/2\sigma_j^2\right], \quad j = 1, 2. \quad (4.46)$$

The mixture density, $h(x)$, has five parameters, θ_1, θ_2, σ_1^2, σ_2^2 and α, which are to be estimated. Pearson used the method of moments, which requires the solution of a ninth-order polynomial; our discussion of the method follows Cohen (1967). Let μ_1', μ_i, $i \geqslant 2$ denote the population moments, and m, m_i, $i \geqslant 2$ the corresponding sample moments, for $h(x)$. Finally put $d_j = \theta_j - \mu_1'$, $j = 1, 2$. We may take $\theta_2 \geqslant \theta_1$ without loss of generality, so that $\theta_1 \leqslant \mu_1' \leqslant \theta_2$ and $d_1 \leqslant 0 \leqslant d_2$. In terms of the unknown parameters the population moments are, therefore,

$$\left.\begin{aligned}
\mu_1' &= \Sigma \, \alpha_j \theta_j, \quad \alpha_1 = \alpha, \, \alpha_2 = 1 - \alpha \\
\mu_2 &= \Sigma \, \alpha_j(\sigma_j^2 + d_j^2) \\
\mu_3 &= \Sigma \, \alpha_j d_j(3\sigma_j^2 + d_j^2) \\
\mu_4 &= \Sigma \, \alpha_j(3\sigma_j^4 + 6\sigma_j^2 \, d_j^2 + d_j^4) \\
\mu_5 &= \Sigma \, \alpha_j d_j(15\sigma_j^4 + 10d_j^2 \, \sigma_j^2 + d_j^4)
\end{aligned}\right\} \quad (4.47)$$

and

the sums being taken over $j = 1, 2$. The equations $\mu_i - m_i = 0$ can then be solved to find the parameter estimates. Equations (4.47) reduce to the single equation

$$\sum_{i=0}^{9} a_i v^i = 0,$$

where $v = d_1 d_2$ and the coefficients are

i	a_i	i	a_i
0	$-24m_3^6$	5	$90k_4^2 + 72k_5 m_3$
1	$-96m_3^4 \, k_4$	6	$36m_3^2$
2	$-63m_3^2 k_4^2 - 72m_3^3 \, m_5$	7	$84k_4$
3	$288m_3^4 - 108m_3 k_4 k_5 + 27k_4^4$	8	0
4	$444k_4 m_3^2 - 18k_5^2$	9	24

where $k_4 = m_4 - 3m_2^2$ and $k_5 = m_5 - 10m_2 m_3$ are sample cumulants. This leads to the estimators

$$\tilde{d}_i = \tfrac{1}{2}\{\tilde{s}_i \pm (s^2 - 4\tilde{v})^{\frac{1}{2}}\},$$

where $\tilde{d}_2$ takes the positive sign, $\tilde{v}$ is the negative real root of the polynomial in v, and $s = v(d_1 + d_2)$ with estimator

$$\tilde{s} = \frac{-8m_3\tilde{v}^3 + 3k_5\tilde{v}^2 + 6m_3 k_4\tilde{v} + 2m_3^3}{\tilde{v}(2\tilde{v}^3 + 3k_4\tilde{v} + 4m_3^2)}.$$

Using these results, we then obtain

$$\tilde{\sigma}_j^2 = \tfrac{1}{3}\,\tilde{d}_j\,(2\tilde{s} - m_3/\tilde{v}) + m_2 - \tilde{d}_j^2; \quad j = 1, 2; \qquad (4.48)$$

$$\tilde{\theta}_j = \tilde{d}_j + m; \quad j = 1, 2; \qquad (4.49)$$

and $$\tilde{\alpha} = \tilde{d}_2/(\tilde{d}_2 - \tilde{d}_1); \qquad (4.50)$$

clearly $0 < \alpha < 1$. It may be possible to improve on these estimates by using them, for example, as the first stage in a minimum chi-square procedure, but little is known about the advantages of such developments. Recently, Robertson and Fryer (1970) have evaluated the bias and accuracy of these moment estimators, using the series expansion approach of Shenton and Bowman (1967); their results suggest that the consistency property of the moment estimators in this context is useless. The conclusion must be that a satisfactory estimation procedure is at present beyond our grasp, for, while the maximum likelihood estimators can be computed iteratively, there is no guarantee of better than a local maximum.

However, various special cases are more amenable to solution:

(a) $\theta_1 = \theta_2 = \theta$ known *or* $\sigma_1 = \sigma_2 = \sigma$ known corresponds to the exponential family situation and can be solved using the methods of sections 4.6 and 4.7.

(b) $\theta_1 = \theta_2 = \theta$, unknown. Tukey (1960) considered the effect on the ML estimator for θ when the sample mean is used, but the population is in fact a mixture of two normals with equal means and different variances. The "contamination" effect of the larger variance can be considerable, so that the median or a linear function of the order statistics is recommended (cf. section 1.9).

(c) $\sigma_1 = \sigma_2 = \sigma$, unknown. This is amenable to ML methods, and Day (1969) considers both the univariate case and its extension to the mixture of two multivariate normals with equal covariance matrices. A combination of moments and frequency moments might also be used. For a review of earlier methods, see Molenaar (1965).

The complexity of the analytical results, and the lack of any ready extension to consider more than two components in the mixture, has led to a variety of graphical methods being proposed. The method given by Harding (1949) is outlined in Exercise 4.9. For more recent developments see Cassie (1962) and Taylor (1965). Such methods extend readily to more than two component populations and appear to give tolerable results whenever their components are reasonably spaced (see Exercise 4.9). However, the subjective nature of the methods makes any assessment of performance difficult.

4.9 The Pearson curves as mixtures

It is well known that the normal mixture, with Pearson's Type V as mixing distribution for σ^2, yields the Student's t, that is

$$h(x; \omega)$$

$$= \frac{1}{\Gamma(k - \frac{1}{2})} \int (2\pi\sigma^2)^{-\frac{1}{2}} \exp\{-\tfrac{1}{2}(x^2 + \omega^2)/\sigma^2\}(\omega^2/2\sigma^2)^{k-(3/2)}\, d(\omega^2/\sigma^2)$$

$$= \{\beta(k - \tfrac{1}{2}, \tfrac{1}{2})\}^{-1}(1 + x^2/\omega^2)^{-k} \tag{4.51}$$

This and other results for the continuous (section 1.2 *et seq.*) and discrete (section 5.1 *et seq.*) Pearson systems are summarised in Table 4.2. We note the link between the first two lines of the table and the

Table 4.2
Mixtures using Pearson curves

Original distribution	Mixing distribution	Mixture
Normal, $f(x; \mu, \sigma^2)$	Type V, $g(\sigma^2; k, \omega^2)$	Student's t, $h(x; \mu, k, \omega^2)$
Type III (gamma), $f(x; k, \lambda)$	Type V, $g(\lambda; m, \mu)$	Type I, $h\left(\dfrac{x}{x + \mu}; \mu, k, m\right)$ Type VI, $h(x; \mu, k, k + m)$
Poisson, $f(r; \lambda)$	Gamma $g(\lambda; k, \mu)$	Negative binomial, $h(r; k, \mu)$
Binomial, $f(r; n, p)$	Type I (beta), $g(p; a, b)$	Beta binomial, $h(r; n, a, b)$
Negative binomial (Pascal), $f(r; k, p)$	Beta, $g(p; a, b)$	Beta–Pascal, $h(r; k, a, b)$

diagram in section 1.2. When f is normal and $g(\sigma^2)$ is gamma, the mixture density is expressible in terms of Hankel functions (Teichroew, 1957).

4.10 Other aspects of mixtures

A problem with somewhat different emphasis, but essentially the same in essence, is the physical mixing of (different sized) particles, there being k types, the ith kind appearing in proportion p_i; e.g. different grades of coal. For analysis of this problem see Buslik (1950), Knott (1967).

An interesting design problem has been investigated by Wilkins (1961). Consider the spraying of crops by helicopter with emissions of spray at (equal) distances/times; the aim is to obtain a uniform spraying of the crop at a given level at minimum cost. The optimal size of emissions and spacing is tabulated by Wilkins when the components are truncated normal distributions.

Daniels (1961) uses a mixture of two geometric components to represent the busy-time distribution in a queueing process; inputs to queues are often tabulated on a priority/non-priority basis making for such a model.

Scale-mixing, where $f(x; \lambda) \equiv f(x; \lambda)$, is discussed by Beale and Mallows (1959) who consider the conditions under which this form is a valid mixture.

Also of interest are the class of "inflated" distributions, with density functions

$$h(x) = \begin{cases} (1 - \alpha)p_0 + \alpha f(0), & x = 0, \\ \alpha f(x), & x > 0, \end{cases} \tag{4.52}$$

notably in dam theory and similar situations. This class was first considered by Aitchison (1955), and Cohen (1960a) has further developed estimators for the parameters. Pandey (1964/5) derived the ML estimators for (4.52) when $f(x)$ is the (truncated) Poisson density function.

EXERCISES

4.1   Use Theorem 4.1 to show that the class of compound Poisson distributions is identifiable. Writing the distribution function (cdf) of the mixture as

$$H(x) = \int_0^\infty \sum_{j \leqslant x} \frac{\lambda^j e^{-\lambda}}{j!} \, dG(\lambda) = \sum_{j \leqslant x} p_j,$$

show that $G^*(\lambda) = 1/p_0 \int_0^\lambda e^{-y} dG(y)$ is also a cdf and has jth moment $p_j j!/p_0$. Use this link between the compound Poisson family and the moments to establish identifiability via the uniqueness of the distribution with such moments.   (Teicher, 1960)

4.2   Let $\alpha_i(\lambda) = \int a_i(x) \, dF(x; \lambda)$ and $\alpha_i(H) = \int a_i(x) \, dH(x)$, where F and H are defined as in equation (4.1) and $\lambda \in \Lambda$. Prove that a necessary and sufficient condition for $\alpha_2(H) \geqslant \alpha_2(\lambda)$, given that $\alpha_1(H) = \alpha_1(\lambda)$, is that α_2 is a convex function of α_1 for $\lambda \in \Lambda$.

Hence show that var $(x_H) \geqslant$ var (x_F) whenever $E(x^2)$ is a convex function of $E(x)$. In particular, show that the class of generalised Poisson distributions (section 6.5) has this property.
                                        (Molenaar and Van Zwet, 1966; Feller, 1943)

4.3   Show that (negative) multinomial mixtures obtained by mixing on the index parameter are identifiable.   (Patil and Bildikar, 1966b)

4.4   Show, for the exponential distribution, that the solutions to (4.26) may not exist and that this occurs with finite probability for all sample sizes.   (Rider, 1961b)

4.5   Given a sample from a population which is a finite mixture of two binomial populations ($n = 20$; p_1, p_2, a unknown), compare the estimates for the unknown parameters given by the method of moments and the moments/linear programming method when $l_1 = 0 \cdot 68$, $l_2 = 0 \cdot 392$, $l_3 = 0 \cdot 2384$.

[*Note*: here $l_1 = m_{(i)}/n^{(i)}$, where $m_{(i)}$ is the ith non-central factorial moment, and $n^{(i)} = n(n - 1) \ldots (n - i + 1)$.]

4.6   For the mixture of exponentials

$$h(x) = \sum_{i=1}^{k} a_i \lambda_i \exp(-\lambda_i x), \quad \Sigma a_i = 1, \quad \lambda_1 < \lambda_2 < \ldots < \lambda_k,$$

show that $h(x)$ is everywhere non-negative if

$$\sum_{i=1}^{j} a_i \lambda_i \geqslant 0, \quad j = 1, 2, \ldots, k.$$

(Bartholomew, 1969)

4.7   If $h(x)$ is the density function of the finite mixture of two normal populations (4.45, 4.46), show that $h(x)$ always has a single stationary point in the range min $(\mu_1, \mu_2) < x < $ max (μ_1, μ_2), implying that the distribution function $H(x)$ has a single inflexion in this range.

4.8   For the finite normal mixture in Exercise 4.7 show that, when the variances are equal, the inflexion of $H(x)$ occurs at $x^* = a\mu_1 + (1 - a)\mu_2$, i.e. at the mean.

4.9   Take the point of inflexion, for the normal mixture in Exercises 4.7 and 4.8, as splitting the population in the proportion $a : 1 - a$. Hence develop a graphical estimation procedure, taking the extremes of the probability plot to represent only one or other component of the mixture. Show that this procedure is "reasonable" provided that "small" values of $\delta = \mu_2 - \mu_1$ do not occur together with $\sigma_2 \gg \sigma_1$ or $\sigma_1 \gg \sigma_2$.   (Harding, 1949)

4.10   A sample of 41 immature copepods was taken; all were the same species and at the same stage of development, but both sexes were present. The size distribution is given in the table. It is known that

82

the females are larger than the males. Estimate the mean size of
the two populations.

Class midpoints (mm)	Frequency	Class midpoints (mm)	Frequency
0·8250	0	0·9375	2
0·8375	2	0·9500	2
0·8500	0	0·9625	3
0·8625	4	0·9750	3
0·8750	2	0·9875	2
0·8875	5	1·0000	3
0·9000	3	1·0125	1
0·9125	2	1·0250	2
0·9250	2	1·0375	2
		—	—
		1·1000	1

(Harding, 1949)

4.11 If X follows the Weibull distribution with density

$$f(x \mid \beta) = \alpha\beta x^{\alpha-1} \exp(-\beta x^{\alpha}), \quad x \geqslant 0, \alpha > 0,$$

show that the compound distribution with compounder

$$g(\beta) = \delta^{\alpha}\beta^{\alpha-1} \exp(-\gamma\beta)/\Gamma(\gamma), \quad \beta \quad 0, \geqslant \gamma, \delta > 0$$

is the Burr distribution given in equation (2.69).

(Dubey, 1968)

CHAPTER 5

THE CLASSICAL DISCRETE DISTRIBUTIONS

5.1 Introduction

It was noted in section 1.1 that Karl Pearson (1895) developed his continuous system from a difference equation, for the density function $\{f_r\}$, of the form

$$\Delta f_{r-1} = \frac{(a - r)f_{r-1}}{b_0 + b_1 r + b_2 r(r - 1)}, \quad r \in T, \tag{5.1}$$

where a and the b_i are parameters and r lies in the range T. This equation was obtained from the hypergeometric distribution. The random variable, R, is called a "lattice random variable", being defined on a regular lattice (of width one); however, any arbitrary width, h, could be used (see Exercise 5.1). Pearson's interests lay with the continuous system and he apparently never developed a discrete analogue. This avenue remained unexplored until Carver (1919) used the difference equation for smoothing actuarial data, but he did not investigate the distributions so defined. However, he later (Carver, 1924) obtained expressions for the parameters in terms of the moments, similar to those for the Pearson system (see Table 5.3, page 97).

Goldberg (1931) and Frisch (1932) used (5.1) in discussions on incomplete moments (see section 5.6), but the first analysis of the density functions obeying (5.1) did not appear until Katz (1946, 1948) and remained unpublished — except for these abstracts — until Katz (1965). The system has been further investigated by Ord (1967a,b). The general class of hypergeometric distributions has been discussed by Davies (1933, 1934), Kemp and Kemp (1956), Sarkadi (1957) and Binet (1968).

The discrete analogue of the exponential family of curves is the class of generalised power series distributions (GPSD), which have been developed in a series of articles by Patil (among others). The best-known discrete distributions — Poisson, binomial, negative binomial (Pascal), logarithmic series — are members of both systems. Despite this overlap we discuss the GPSD separately in Chapter 6.

The division between "classical" and "contagious" distributions is by no means strict (e.g. the ubiquitous negative binomial), but the assumptions underlying the models differ, so we reserve the discussion of contagious models for Chapter 6 also. Various links between the distributions are also explored in section 7.2.

In the current chapter we also discuss the series expansion forms.

5.2 The difference equation system

We may solve (5.1) as we did the differential equation (1.4) to obtain the set of distributions in the system (see the discussion of section 1.3). The results appear in Table 5.1, pages 86–7.

The numbering of the main types is based on the roots of the quadratic in the denominator of (5.1). The breakdown is:

Type I: one root zero, the other non-zero, finite range (excluding Type I(e)),

Type VI: one root zero, the other negative, infinite range;

Type IV: roots imaginary.

This differs from the basis of the numbering system used for Pearson curves, but not violently so. Otherwise, when possible, we have numbered the distributions in the system to correspond to Elderton's notation for Pearson curves. The letters "d" for discrete, "c" for continuous, have been appended to the type number (e.g. Type III(d)) whenever this is not clear from the context.

The kappa criterion used to distinguish the distributions is based on the roots of the quadratic in the denominator

$$\Delta f_{r-1} = \frac{(a - r) f_r}{b_0 + a + (b_1 - 1)r + b_2 r(r - 1)} \tag{5.2}$$

rather than (5.1); the two equations are equivalent, but (5.2) is preferred for the present purpose as $b_0 = 0$ in many cases (Types I and VI above). The criterion used is therefore

$$\kappa = (b_1 - b_2 - 1)^2 / 4b_2(a + b_0), \tag{5.3}$$

where the roots of the denominator of (5.4) are real unless $0 < \kappa < 1$. This criterion is supplemented by the index of dispersion, or "clumping",

$$I = \mu_2 / \mu_1', \tag{5.4}$$

which is of value when the range of the random variable has a finite lower end-point, which may be regarded as the origin.

5.3 The densities contained in the system

Although, formally, the distributions generated by (5.1) are special cases of the solution to that difference equation, interest in the system is sustained, as for the original Pearson curves, by the value of these forms to statistics. We now briefly summarise how the different forms arise in practice.

I(a)   Direct attribute sampling without replacement from a finite population (population size N, number with the attribute $M = Np$, sample size n).

I(b)   Inverse attribute sampling until the kth success or until the population is exhausted, without replacement from a finite population; a binomial population with beta mixing (Skellam 1948, 1949); binomial sampling (i.e. with replacement) given prior information expressed as a beta distribution for $p = M/N$ (Weiler 1965; Raiffa and Schlaifer 1961). This type includes the rectangular distribution, $f_r = N^{-1}$, $r \in [0, N - 1]$ as a special case.

I(e)   A mathematical accident caused by the discrete nature of r.

I(U),   The U-shaped distributions have no sampling interpretation,
III(U) and data of such a form are often better described by a mixture of two or more distributions, or by sampling on a Markov chain (Gabriel 1959). However, the "inverted" binomial Type III(U) may be regarded as describing a stochastic process in which there is a "bandwagon effect", units tending to join the larger group (see Exercise 5.3).

VI   The negative binomial (or Pascal) distribution with beta mixing. Inverse binomial sampling with prior information expressed in beta form (Raiffa and Schlaifer, 1961). In the context of accident proneness theory this is known as the generalised Waring distribution (Irwin 1965, 1968). The J-shaped distributions which occur when $N - M = 1$ (see Table 5.1) have been considered by Marlow (1966). When $N - M = k = 1$, $f_r \propto \binom{N+r-1}{r}^{-1}$, the limiting Yule distribution for the simple birth-and-death process. Where large values of r are of interest, $f_r \propto r^{-N+1}$, approximately, which is the discrete Pareto form (Seal, 1952, 1953; Simon, 1955; Rider, 1965).

IV, V   As in the Pearson curve system, these have no natural interpretation; k is taken as a positive integer.

VII   The discrete Student's t, including the discrete Cauchy. The most interesting feature of this type is its peculiar form of asymmetry (see Exercise 5.4).

Table 5.1 *Distributions contained in the system, with range,*

Type (d)	Name	Density
I(a)	Hypergeometric	$\binom{M}{r}\binom{N-M}{n-r}\Big/\binom{N}{n}$
I(b)	Negative hyper-geometric or beta binomial	$\binom{k+r-1}{r}\binom{N-k-r}{M-r}\Big/\binom{N}{M}$
I(e)	–	$\binom{A}{r}\binom{C}{B-r}\Big/\binom{A+C}{B}$
I(U)	–	$\alpha\left\{\binom{A}{C+r}\binom{B}{D-r}\right\}^{-1}$
VI	Beta–Pascal	$\dfrac{A}{(k+A)}\binom{k+r-1}{r}\binom{A+B-1}{A}\Big/\binom{k+A+B+r-1}{k+A}$
IV	–	$\alpha Q(r,a,d)/Q(r,k+a,b), r>0$; similar expression for $r<0$
II(a) II(b) II(U)	As for Type I(.)	
V	–	As Type IV, but $b=0$
III(B)	Binomial	$\binom{n}{r}p^r(1-p)^{n-r}$
III(N)	Negative bi-nomial or Pascal	$\binom{k+r-1}{r}p^k(1-p)^r$
III(P)	Poisson	$e^{-m}m^r/r!$
VII	Discrete Student's t	$\alpha\left[\displaystyle\prod_{j=1}^{k}\{(j+r+a)^2+b^2\}\right]^{-1}$

Notes: (1) $Q(r,a,d)=(a^2+d^2)\{(a+1)^2+d^2\}\ldots\{(a+r)^2+d^2\}$.

(2) Except for random variable r, all symbols are parameters and

(3) α is a constant such that the total measure is 1.

density function and criteria values (reproduced from Ord, 1967b)

Criteria	Range	Comments
$l < 1, \kappa > 1$	$[0, m]$ $m = \min(n, M)$	J- or bell-shaped
$\kappa < 0$	$[0, M]$	J- or bell-shaped
$\kappa > 1$	$[0, \infty)$	A, C non-integer, but have the same integral part
$\kappa > 1$	$[0, n], n < D$	U-shaped
$l > 1, \kappa > 1$	$\begin{cases} [0, \infty) & A = M - 1 \\ & B = N - M \end{cases}$	J- or bell-shaped
$0 < \kappa < 1$	$(-\infty, \infty)$	k a positive integer; bell-shaped
$\begin{cases} l < 1, \kappa = 1 \\ \kappa = 0 \\ \kappa = 1 \end{cases}$	As Type I(.)	Symmetric forms of Type I(.)
$\kappa = 0$	$[0, \infty)$ or $(-\infty, \infty)$	Limiting form of IV
$l < 1, \kappa \to \infty$	$[0, n]$	Limiting form of I(a), I(b)
$l > 1, \kappa \to \infty$	$[0, \infty)$	Limiting form of I(b), VI
$l = 1, \kappa \to \infty$	$[0, \infty)$	Limiting form of III(B), III(N)
$0 < \kappa < 1$	$(-\infty, \infty)$	"Nearly" symmetric form of IV

$p = 1 - q$. The usual notation is followed where possible.

III   The Type III distributions need no introduction, but we recall that the negative binomial appears in four different contexts (Anscombe, 1949), namely inverse binomial sampling; immigration-birth-death process; Poisson sampling with gamma mixing; and Poisson distribution of "colonies", the size of which is described by the logarithmic series distribution. Other processes which yield the negative binomial are discussed by Boswell and Patil (1970). For a review of the properties of the NBD, see Bartko (1961).

For a comprehensive account of the Poisson distribution, its extensions and applications, see Haight (1967).

For the "real roots" distributions, excepting the U-shaped curves, $b_0 = 0$ in (5.1), while for the Type III densities $b_2 = 0$. If we keep $b_2 = 0$, but allow b_0 to take non-zero values, then provided $b_0 + b_1 j > 0$ the equation

$$\Delta f_{r-1} = (a - r)f_{r-1}/(b_0 + b_1 r), \quad r = j, j + 1, \ldots, \quad (5.5)$$

first discussed by Katz (1946), yields a system of "hyper" distributions. The hyper-Poisson ($b_1 = 1$) takes the form

$$f_r = f_0 \lambda^r \frac{\Gamma(1 + b)}{\Gamma(1 + b - r)}, \quad r \geqslant j \geqslant 1, \quad (5.6)$$

investigated by Bardwell and Crow (1964, 1965) and Staff (1964, 1967) – the latter using the term "displaced" Poisson. The f_r in equation (5.6) are the terms of a confluent hypergeometric function, tables of which are available (Slater, 1960).

According to $b_1 <$ or > 1 in equation (5.5), we have the three-parameter hyperbinomial and hyper-Pascal families. Such forms may be of value when systematic discrepancies are found using the usual attribute sampling models.

When $b_0 = 0$ or $b_1 = 1$ in equation (5.5), we note that the resulting density is of the power-series form (see section 6.4).

5.4   The Pólya urn scheme

All the "real roots" distributions may be interpreted in an attribute sampling context. These distributions may therefore be regarded as arising from the Pólya urn scheme. This is formulated as follows:

Consider an urn containing M black and $N - M$ white balls. A ball is drawn and its colour noted. The ball is replaced and c further balls of the same colour added to the urn. The next drawing is then made.

Any stopping rule may be specified for the sampling process, e.g. stop after n drawings have been made, stop after k black balls have been drawn.

For direct sampling, the probability of selecting r black balls in n drawings is (Bosch, 1963)

$$f_r = \binom{n}{r} (M)^{(r,c)}(N - M)^{(n-r,c)}/(N)^{(n,c)}, \tag{5.7}$$

while for the inverse case

$$f_r = \binom{k + r - 1}{r} (M)^{(k,c)}(N - M)^{(r,c)}/(N)^{(k+r,c)}, \tag{5.8}$$

where $\qquad (A)^{(x,c)} = A(A + c) \ldots \{A + (x - 1)c\}.$

For example, (5.7) and (5.8) yield the following sampling schemes:

$c = -1$, without replacement: direct — hypergeometric; inverse — negative hypergeometric (beta–binomial).

$c = 0$, with replacement: direct — binomial; inverse — negative binomial.

$c = +1$, with replacement and addition of a like unit (binary fission): direct — beta–binomial; inverse — beta–Pascal.

The various distributions and their inter-relations are summarised in Fig. 5.1.

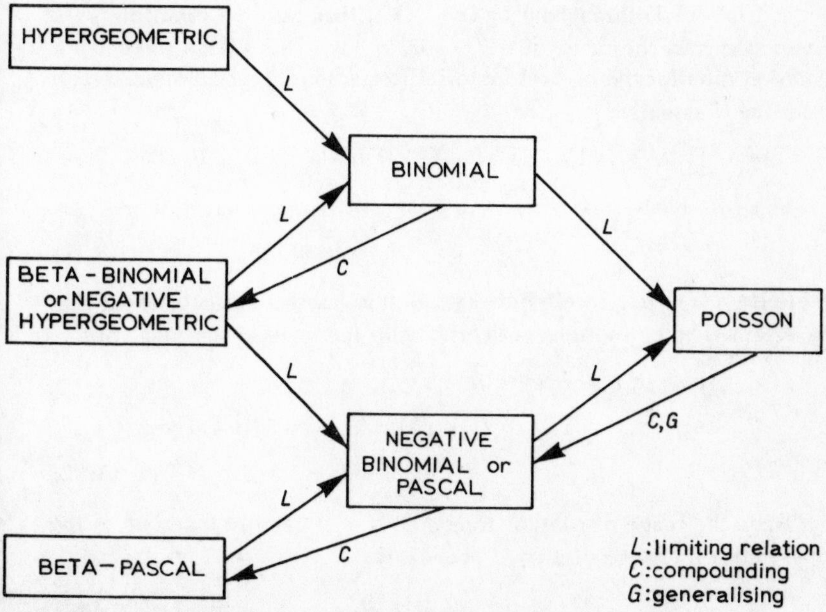

Fig.5.1   The relationships between the classical discrete distributions

The jth non-central factorial moments are

$$\mu'_{(j)} = n^{(j)} (M)^{(j,\,c)}/(N)^{(j,\,c)} \tag{5.9}$$

$$= (k)^{(j,\,1)} (M)^{(j,\,c)}/(N)^{(j,\,c)} \tag{5.10}$$

for the direct and inverse schemes respectively.

These results may also be obtained from (5.11) below. The basic urn scheme may be extended by allowing the number of balls added after the ith drawing to vary according to the number of the drawing and the colour of the ball drawn (see Freedman, 1965, for example). This extended model may be regarded as a general birth-and-death process and has been used by several authors (Cernuischi and Castagnetto, 1946; D.G. Kendall, 1948a, b; Woodbury, 1949, being some of the earlier studies of importance).

5.5 Moments

Recurrence relations for the moments of certain distributions (notably the Poisson, binomial, and negative binomial) have been derived by many authors — see, for example, Romanovsky (1923) and Riordan (1937). We now derive a general relationship for the family described by (5.1).

Multiply (5.1) throughout by $(r - \mu)^{(j)}$, then sum the resulting expression over the range of r, $T = [u, v]$ say. This yields a recurrence relation for the central factorial moments $\mu_{(j)}$ (about mean μ), assuming they exist,

$$\{(j + 2)b_2 - 1\} \, \mu_{(j+1)} + \{(j + 1)(b_1 + 2(j + \mu)b_2) + a - \mu - 2j - 1\} \, \mu_{(j)} +$$
$$+ j\{b_0 + b_1(j + \mu) + b_2(j + \mu)^{(2)} + a - \mu - j\} \, \mu_{(j-1)} - E_{j+1} = 0$$

$$\text{for } j = 0, 1, \ldots; \tag{5.11}$$

where $\mu_{(-1)} = 0$, $\mu_{(0)} = 1$. Putting $\mu = 0$ gives the moments about the the origin. The correction terms, E_{j+1}, for the ends of the range are

$$E_{j+1} = - \{b_0 + b_1 u + b_2 u^{(2)}\} \, (u - \mu)^{(j)} f_u +$$
$$+ \{b_0 + b_1(v + 1) + b_2(v + 1)^{(2)} + a - v - 1\}(v + 1 - \mu)^{(j)} f_v. \tag{5.12}$$

When the range of r is not truncated it will usually take one of the forms, after a change of sign if necessary,

$$(-\infty, \infty), \; [0, \infty), \; [0, v].$$

Where one or both terminals are infinite, $\mu_{(j)}$ will exist, provided that

$$\lim_{r \to \pm\infty} r^{j+1} f_r = 0 \quad \text{if } b_2 \neq 0,$$

$$\lim_{r \to \infty} r^j f_r = 0 \quad \text{if } b_2 = 0, \text{ when } r \in [0, \infty).$$

If $u = 0$, the first term in (5.12) reduces to $-b_0 f_0 (-\mu)^{(j)}$, and is zero for $\mu \to -\infty$.

Further, for the "real roots" distributions with a finite range $[u, v]$ it may be shown that

$$a + b_0 + (v + 1)(b_1 + b_2 v - 1) = 0. \tag{5.13}$$

For the "real roots" distributions, when $b_0 = 0$, $u = 0$ and (5.13) holds, the first four moments are

$$\mu_1' = (b_1 + a - 1)/(1 - 2b_2) \tag{5.14a}$$

$$\mu_2 = (b_1 + a - 1)\{ab_2 + (1 - b_2)(b_1 - 2b_2)\}/\{(1 - 2b_2)^2(1 - 3b_2)\} \tag{5.14b}$$

$$\mu_3 = \mu_2\{4b_2(a + 1 - b_1 - b_2) + 4b_1 - 3\}/\{(1 - 2b_2)(1 - 4b_2)\} \tag{5.14c}$$

$$\mu_4 = 3\mu_2^2 + \frac{\mu_2}{(1 - 4b_2)(1 - 5b_2)}[(1 - b_2)(1 - 2b_2) +$$
$$+ 6\{(b_1^2 - b_2^2) + (b_2 - b_1) - 2b_2(a + b_1 - 1)\} +$$
$$+ b_2\mu_2\{1 - 2b_2(1 - 3b_2)\}^{-1}\}]. \tag{5.14d}$$

Although the general form of (5.11) is rather cumbersome it reduces considerably in special cases.

Example 5.1

The Poisson density, $f_r = e^{-\lambda} \lambda^r/r!$ has $a = \lambda$, $b_1 = 1$, $b_0 = b_2 = 0$.

For the non-central factorial moments, $\mu_{(j)}'$, (5.11) reduces to

$$j\lambda\mu_{(j-1)}' + (\lambda - j)\mu_{(j)}' = \mu_{(j+1)}'$$

or $\quad \mu_{(j+1)}' - \lambda\mu_{(j)}' = j(\mu_{(j)}' - \lambda\mu_{(j-1)}')$

$$= \cdots\cdots$$

$$= \mu_{(1)}' - \lambda = 0$$

that is, $\quad \mu_{(j+1)}' = \lambda\mu_{(j)}' = \lambda^{j+1}$.

Example 5.2

For the hyper-Poisson family in (5.6), $b_2 = 0$, $b_1 = 1$, $b_0 = b$, $a = \lambda - b$, so that

$$\mu'_{(1)} = \mu = \lambda - b + bf_0,$$

$$\mu_{(j+1)} = \lambda j \mu_{(j-1)} - (j + bf_0)\mu_{(j)} + bf_0(-\mu)^{(j)}.$$

Using this relation, we find that $\mu_2 = \lambda - \mu b f_0$. Hence $I = \mu_2/\mu'_1$ $= 1 + b(1 - f_0 - f_0\mu)/\mu$. It is readily shown that $f_0 < (1 + \mu)^{-1}$ for finite b, so that

$$I > 1 \text{ when } b > 0 \text{ (super-Poisson)}$$
$$< 1 \text{ when } b < 0 \text{ (sub-Poisson)}.$$

5.6 Properties of distributions in the system

For those distributions in the system whose density function may be represented as the terms of a hypergeometric series, the probability generating function may be written as

$$f_0\{F(\alpha_1, \alpha_2; (b_1 + 1 - b_2)/b_2; z)\}, \qquad (5.15)$$

where α_1, α_2 are the roots of

$$\alpha^2 b_2 + (b_1 - 1 - b_2)\alpha + a = 0.$$

The characteristic function is given by replacing z by $\exp(it)$, $i = \sqrt{-1}$. Other properties obeyed by this class are now listed.

Property D1. Turning-points. From (5.1) it is evident that all distributions in the system are either J-shaped or unimodal (single maximum or minimum). If a turning-point exists, a maximum for f_r occurs at $r = r_0$ where

$$(a - r)\{b_0 + b_1 r + b_2 r^{(2)}\}^{-1} \leqslant 0 \text{ for } r > r_0$$
$$\geqslant 0 \text{ for } r \leqslant r_0 \qquad (5.16)$$

where the value of the density function may be the same for two or more adjacent values of r (e.g. the rectangular distribution). If the turning-point is a minimum the inequalities in (5.16) should be reversed. Janardan and Patil (1970) have determined the modal value for various members of this family and certain other discrete distributions.

Property D2. Points of inflexion. An inflexion of f_r occurs at that value of r for which $\Delta^2 f_{r-1}$ changes sign. The discrete distributions have 0, 1 or 2 inflexions occurring at the smallest integers greater than the roots of

$$r^2(1 + b_2) - r(2a - 2 + 2ab_2 - b_2) + a^2 - 2a - b_0 - b_1 a = 0,$$
$$(5.17)$$

which lie within the range of r values. We restrict the discussion to the case where the mode and inflexions are uniquely defined. Suppose two points r_L, r_U exist, corresponding to the roots of (5.17), $r_L - b$, $r_U - c$, $0 \leqslant b, c < 1$, while the mode occurs at r_0, (5.16) being zero at $r_0 - h$, $0 \leqslant h < 1$. Then the inflexions are approximately equidistant from the mode, in the sense that

$$(r_0 - h) - (r_L - b) = (r_U - c) - (r_0 - h).$$

The correspondence with the exact result for Pearson curves (Property P5 in section 1.4) is obvious.

Property D3. Mean deviation (Kamat 1966b; Ord 1967a). Define the mean deviation about the mean μ as

$$e_1 = E(|R - \mu|),$$
$$= \sum_{r=u}^{m} (\mu - r)f_r + \sum_{r=m+1}^{v} (r - \mu)f_r, \qquad (5.18)$$

where $m = [\mu]$, the greatest integer less than or equal to μ. Using (5.1) and the identity $E(R - \mu) \equiv 0$,

$$(1 - 2b_2)e_1 = 2[f_m\{b_0 + mb_1 + m^2 b_2 + (\mu - m)(1 - b_2)\} - f_u\{b_0(1 - S) + b_1 u + 2b_2 u^2\}], \qquad (5.19)$$

where $S = f_u + \ldots + f_m$, and either (5.13) holds or v is infinite. When $u = 0$ or $u \to -\infty$, (5.19) simplifies considerably. If, further, $b_0 = 0$, we have

$$(1 - 2b_2)e_1 = 2f_m\{mb_1 + (\mu - m)(1 - 2b_2) + m(m - 1)b_2\}. (5.20)$$

For μ an integer, $m = \mu$, while (5.11) yields

$$(1 - 3b_2)\mu_2 = \mu b_1 + \mu(\mu - 1)b_2,$$

so that
$$e_1 = 2\mu_2 f_m(1 - 3b_2)/(1 - 2b_2). \qquad (5.21)$$

Thus the Johnson property

$$e_1 = 2\mu_2 f_m$$

holds only when $b_2 = 0$; that is for the Poisson, binomial and negative binomial. The values of e_1 for certain of the discrete forms are listed in Table 5.2 (Ramasubban, 1958).

Table 5.2

Mean difference (Δ_1) *and mean deviation* (e_1) *for selected discrete distributions (when* $m = \mu$)

Distribution	e_1	Δ_1
Binomial	$2\mu_2 f_m$	$2\mu_2\, F(-n+1,\ \tfrac{1}{2};\ 2;\ 4pq)$
Negative binomial	$2\mu_2 f_m$	$2\mu_2\, F(k+1,\ \tfrac{1}{2};\ 2;\ -4q/p)$
Poisson	$2\mu_2 f_m$	$2\mu_2\, \phi(\tfrac{1}{2};\ 2;\ -4\lambda)$
Hypergeometric	$2\mu_2\, f_m(1 + N^{-1})$	
Beta–binomial	$2\mu_2\, f_m\{1 + (M+1)^{-1}\}$	
Beta–Pascal	$2\mu_2\, f_m\{1 - (M-1)^{-1}\}$	
Types IV, VII	$2\mu_2\, f_m\{1 - (2k)^{-1}\}$	
Truncated Poisson	$2mf_m - 2uf_u$	

Note: F is the hypergeometric function and ϕ the confluent hypergeometric.

Property D4. Mean difference (Katti, 1960; Ord, 1967a). The generalised mean difference is

$$\Delta_k = \sum_i \sum_j |i - j|^k f_i f_j. \tag{5.22}$$

When k is even, $= 2p$ say, $\Delta_{2p} = 2\mu_{2p}$. For odd k, only the mean difference, Δ_1, is of major interest. For the evaluation of higher Δ_k in special cases, see Ramasubban (1959).

If $G(z)$ is the generating function of *any* discrete distribution, then

$$\Delta_1 = \left[P\, \frac{d}{dz} \{G(z)\, G(z^{-1})\} \right]_{z=1} \tag{5.23}$$

where the operator P denotes summation over all terms of non-negative order in z.

Property D5. Incomplete moments (Guldberg, 1931; Kamat, 1965). If r has the range $[0, n]$, $n \leqslant \infty$, we define the jth incomplete moment about a as

$$\alpha_j(u, v) \equiv \alpha_j = \sum_{r=u}^{v} (r - a)^j f_r, \quad 0 \leqslant u \leqslant r \leqslant v \leqslant n, \tag{5.24}$$

where a is given by (5.1). Provided the density function satisfies (5.1) and condition (5.13),

$$\{1 - (j - 1)B_2\}a_j = (u - a)^{j-1}f_{u-1}Q_u + (B_1a_{j-1} + B_0a_{j-2})(j - 1) +$$

$$+ \sum_{i=0}^{j-3} \binom{j-1}{i}(B_2a_{j+2} + B_1a_{j+1} + B_0a_j), \qquad (5.25)$$

where $\quad Q_r = (r - a)^2 B_2 + (r - a)B_1 + B_0,$

$\qquad\qquad B_2 = b_2,$

$\qquad\qquad B_1 = b_1 - b_2 - 1 + 2ab_2,$

and $\qquad B_0 = b_0 + b_1a + b_2a(a - 1).$

((5.1) may be rewritten as $(r - a)f_r = f_{r-1} Q_r - f_r Q_{r+1})$. This relationship gives that of Frisch (1932), for the binomial, as a special case. Further, Romanovsky's (1923) result for the binomial may be written in the form

$$a_{j+1} = \omega b_1 \frac{da_j}{d\omega} + \sigma^2 a_{j-1}, \qquad (5.26)$$

where $\omega = p, q/p, \lambda;\ b_1 = q, p^{-1}, 1$, as in (5.1), for the binomial, negative binomial and Poisson respectively. σ^2 is the variance of the *complete* distribution in all cases.

5.7 Fitting by moments

When the correction terms in (5.11) disappear it is straightforward to solve for the j parameters in terms of the first j moments. This was done (i) for $j = 4$ (Carver, 1923); (ii) for $j = 3$, when $b_0 = 0$ (Ord, 1967b), using (5.14). The expressions appear in Table 5.3, the Pearson curve results being included for comparison. To select a particular distribution given the theoretical or sample moments, we could use, as for the Pearson curves, the β_1, β_2 chart. The β_1, β_2 regions for the different distributions were first obtained by Davies (1933) by considering the functions

$$\theta = \beta_2 - \beta_1 - 3 \qquad (\theta = 0 \text{ for Poisson})$$

and $\qquad \phi = 2\beta_2 - 3\beta_1 - 6 \qquad (\phi = 0 \text{ for Type IIIc}).$

Using (5.14) we may establish the boundaries for the different distributions shown in Fig. 5.2. By taking the lattice to be of variable width, h, rather than a fixed value ($h = 1$), the Pearson curves are obtained as limiting forms as $h \to 0$. Also, for small b, d we may show

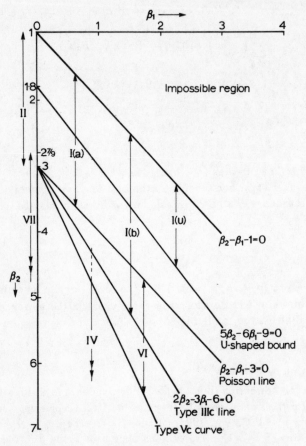

Fig.5.2   The β_1, β_2 chart for the discrete Pearson system
(Adapted from Ord, 1967b)

that for Types IV, VII, for given k,

$$\min \beta_1 = 0 \quad \text{and} \quad \min \beta_2 = 3 - \frac{2(2-k)}{k(2k-3)},$$

so that $\min \beta_2 = 2\frac{7}{9}$ when $k = 3$.

It is conjectured (Ord, 1967a) that the U-shaped forms satisfy

$$5\beta_2 - 6\beta_1 - 9 + \frac{1}{\mu_2} \leqslant 0, \tag{5.27}$$

as this is an equality for the rectangular distribution, and is always

Table 5.3

Values of the parameters in terms of the first three (or four) moments
(Reproduced from Ord, 1967b)

Parameter	Pearson	Discrete with range $(-\infty, \infty)$	Discrete with range $[0, N]$
a	$-(\mu_3/\mu_2)(\beta_2 + 3)D_P$	$\frac{1}{2} - (\mu_3/\mu_2)(\beta_2 + 3 - 1/\mu_2)D_F$	$(1 - 2b_2)\mu + 1 - b_1$
b_0	$\mu_2(4\beta_2 - 3\beta_1)/D_P$	$\mu_2(4\beta_2 - 3\beta_1 - 1/\mu_2)/D_F$	0
b_1	$-a$	$1 - a$	$\{\mu_2 - b_2(3\mu_2 + \mu^2 - \mu)\}/\mu$
b_2	$(2\beta_2 - 3\beta_1 - 6)/D_P$	$(2\beta_2 - 3\beta_1 - 6 + 1/\mu_2)/D_F$	$\{\mu(\mu_3 + \mu_2) - 2\mu_2^2\}/D_G$

$D_P = 2(5\beta_2 - 6\beta_1 - 9)$, $\quad D_F = 2(5\beta_2 - 6\beta_1 - 9 + 1/\mu_2)$, $\quad D_G = 4\mu\mu_3 + 2\mu_2(\mu + \mu^2 - 3\mu_2)$.

satisfied for density functions of the form

$$f_r = M^{-1}, \quad r = -M, \ldots, -1,$$
$$= (N + 1)^{-1}, \quad r = 0, 1, \ldots, N.$$

Further, dropping the term h/μ_2, as $h \to 0$, we obtain the Pearson U-curve boundary.

Unfortunately, the regions in the β_1, β_2 plane for the different densities overlap considerably, so the chart's main use lies in distinguishing Types IV and VII. The other densities listed in Table 5.1 have, however, range $[0, v]$, $v \leqslant \infty$ and $b_0 = 0$. As is seen from Table 5.3, only three moments are required to define the parameters. We take

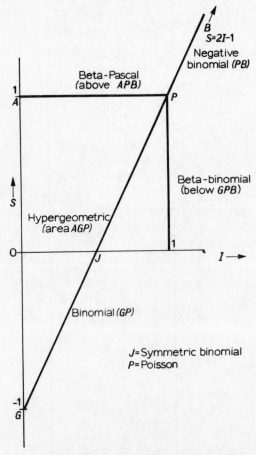

Fig. 5.3   The I, S chart for the discrete Pearson system
(Adapted from Ord, 1967b)

advantage of this situation by employing the I, S chart (Fig. 5.3), where $I = \mu_2/\mu_1'$ is the index of dispersion and $S = \mu_3/\mu_2$.

Omitting the pathological Type I(e) from consideration, we see that each distribution occupies a uniquely defined part of the I, S region. Further, since only the first three moments are used to define I and S, the variances of the selection criteria will be much smaller.

The central line in Fig. 5.3, $S = 2I - 1$, defines the binomial, Poisson or negative binomial as $I <, =, > 1$. The other regions follow from analysis (Ord, 1967b) of

$$\delta = S - 2I + 1. \tag{5.28}$$

5.8 Efficiency of moment estimators

Asymptotically the maximum likelihood estimators (MLE) are fully efficient. For the Type IIId distributions, which are all generalised power series distributions (GPSD), the mean is the MLE for a function of the single parameter (p or λ). For the hypergeometric series distributions there are three parameters, some of which may be known from prior considerations.

Only M unknown

Take, for example, Type I(a) with

$$f_r = \binom{M}{r}\binom{N-M}{n-r}\Big/\binom{N}{n}. \tag{5.29}$$

Evaluating $\Delta_M\{\log\ \text{likelihood}\}$, we find that the MLE for M, given a single sample of size n, is the largest integer value $\hat{M}$ such that

$$(n - r)(\hat{M} + 1) > (N - \hat{M})r, \tag{5.30}$$

which differs by at most 1 from the sample mean estimator. M is the integer nearest to Nr/n.

Similar expressions can be found for the other hypergeometric series distributions (Finucan, 1964).

M and N unknown

The ML equations become very tedious to solve, but it is instructive to explore the efficiency of the moment estimators in this case. These results are an extension of work by Shenton (1950).

We measure efficiency, E, by

$$E = \lim_{m \to \infty} \frac{\text{variance (MLE)}}{\text{variance (other estimate)}}, \tag{5.31}$$

where m samples are taken. E is extended to cover multiparameter estimation by using the generalised variance; that is

$$E = \lim_{m \to \infty} |V(\text{ML})| / |V(\text{other})| . \qquad (5.32)$$

Let the density function of a distribution based on the hypergeometric series be given by the terms of

$$f_0 F(a, b; c; 1) = f_0 \left(1 + \frac{ab}{c} + \frac{ab(a + 1)(b + 1)}{c(c + 1)2!} + \ldots \right), \qquad (5.33)$$

typified by the triple (a, b, e), where $e = c - a - b - 1$. Then, using (5.32), it may be established that

$$E^{-1} = 1 + \frac{4e(a + 2)(e - 2)(e - 4)(e + 2b + 1)^2}{3(b + 2)(e + b - 1)(e + a - 1)(e - 1)^4} +$$

$$+ \frac{2(a + 1)^2 (b + 1)(e + 1)(e + b)(e - 4)(e - 3)^2}{3e^2(b + 2)(e + b - 1)(e - 2)^4 (e + a)^{(2)}} +$$

$$+ \frac{3e(a + 2)(a + 3)(e - 6)\{(b + 3)^{(3)} + (e + b)^{(3)}\}}{(b + 3)^{(2)} (e + b - 1)^{(2)} (e - 1)^3 (e - 2)^3 (e + a - 1)^{(2)}}$$

$$+ \ldots . \qquad (5.34)$$

where $u^{(j)} = u(u - 1) \ldots (u - j + 1)$.

We now discuss Types I(b), I(a) and VI, using the notation of Table 5.1.

I(b) — Negative hypergeometric (Shenton, 1950) with triple $(k, M - N, -M - 1)$:

(i) if $N - M$ is small, M not small,

$$E^{-1} \sim 1 + \frac{4(k + 2)}{3(M + k)(N - M - 2)}, \qquad (5.35)$$

which rarely falls below 70 per cent.

(ii) if M is small (near k), E may be very low.

(iii) if $M, N - M$ are both large, the distribution approaches the negative binomial and $p = M/N$ becomes the parameter of major interest. Efficiency will therefore be high.

(iv) if $M, N - M$ are both small — as may be the case for the urn sampling scheme when $c = +1$ (see section 5.4) — then

$$E^{-1} \sim 1 + \frac{7(k - 2)(k - 3)}{18(k + 2)(k + 3)}, \qquad (5.36)$$

which yields $E \geqslant 80$ per cent for $k \leqslant 20$, covering most cases of practical interest.

I(a) — Hypergeometric $(-n, -M, N)$:

(i) if M or $N - M$ is very small, N not large, then E is low.

(ii) when $M \simeq N - M$, $n/N < 0.5$ say, then

$$E^{-1} \sim 1 + \frac{16(n - 2)}{3(N - n - 1)(N - 1)^4}, \qquad (5.37)$$

so that $E \geqslant 90$ per cent.

(iii) when N, M are large, the limiting binomial parameter, $p = M/N$, is efficiently estimated by the mean.

VI — Beta–Pascal $(k, B, A - 1)$:

(i) when A, B are large the efficiency is high; otherwise E may be low.

In summary, it can be said that when the configuration of parameters is such that the distribution is highly skew, the moments estimators are usually of low efficiency.

Three unknowns

No study has been made of the efficiency of moment estimators in the general case; but by analogy with the two-moments method for the negative binomial we must expect poor results.

The analysis here is based on asymptotic results. Much more could be discovered by numerical small-sample studies, and even for the best-known distributions much more work needs to be done.

5.9 Other methods

In addition to the methods of moments and of maximum likelihood, we could use frequency moments or the values of particular observed frequencies. A further alternative, the minimum chi-squared method, is applied to certain contagious distributions in Chapter 6.

The frequency moments (see section 1.10) are most useful when the population moments fail to exist; for the discrete Cauchy (Type VII(d) with $k = 0$), the second frequency moment is asymptotically fully efficient for b (see Exercise 5.12). For Type VII, the second frequency moment is likely to be fairly efficient for $k \leqslant 10$ say, beyond which the usual moment estimators are acceptable. As the frequency moments are location independent, an estimator for the location parameter is required. The median will usually be fairly efficient for a.

Using the observed relative frequencies for given values of r often provides a source of reasonably efficient, easily used estimators.

Example 5.3

The Type III(U) density may be written as

$$f_r \propto (C + r)! \, (D - r)! \, p^r q^{n-r}$$

$$r = 0, 1, \ldots, n; \, p + q = 1; \, C > 0, D > n.$$

When C, D are known (see Exercise 5.3) the sample mean is fully efficient for p as the distribution is a GPSD (see section 6.4). However, the population mean is

$$\mu = q(1 - f_0)(D + 1) - p(C + 1)(1 - f_n) + npf_n,$$

which is extremely difficult to solve except for very small n. A convenient method would be, therefore, to use the sample mean with the observed values for f_0 and f_n.

When C and D are also unknown, we could use the first three sample moments and the terminal observed relative frequencies. As $r = 0$ and $r = n$ are likely to account for a high proportion of the observed values, the use of the observed relative frequencies should not seriously reduce the efficiency.

Provided the values of r are chosen so that the relative frequencies f_r are expected to be large (as in Example 5.3), this approach often yields estimators of surprisingly high efficiency. The method is of particular value for truncated distributions (see section 5.12). We now turn to a method of using the observed f_r which enables the selection of a distribution to be carried out graphically (Ord, 1967c).

5.10 A plotting method

For the hypergeometric series distributions and their special cases, we may rewrite (5.1) as

$$u_r = rf_r/f_{r-1} = r + (a - r)(b_1 + b_2 r)^{-1}, \tag{5.38}$$

reducing to

$$u_r = r + (a - r)/b_1 \equiv c_0 + c_1 r \tag{5.39}$$

for the Type III distributions in the group. The expressions for c_0 and c_1 are given in Table 5.4 (that for the binomial was first obtained by Dubey, 1966).

By considering $b_2 \neq 0$, the deviations of the hypergeometric distributions from the lines in Table 5.4 may be found; these are summarised

Table 5.4

The intercept and slope of the plot rf_r/f_{r-1} against r for certain discrete distributions

Distribution	Intercept (c_0)	Slope (c_1)
Binomial	$(n + 1)p/q$	$-p/q$
Negative binomial	$(k - 1)q$	q
Poisson	λ	0
Logarithmic [see (5.43)]	$-q$	q
Rectangular	0	1

in Table 5.5, and typical graphs are given in Fig. 5.4. Further, for the hyper-Poisson,

$$u_r = \lambda - \lambda b(b + r)^{-1}, \tag{5.40}$$

yielding plots above and below the Poisson line for $b >$ or < 0.

Table 5.5

The beginning and end-points of the hypergeometric distribution u_r curves compared with the binomial and negative binomial lines

Distribution	Beginning	End
Hypergeometric	Above binomial	Below binomial
Beta–binomial	Below binomial	Above binomial
	Above negative binomial	Below negative binomial
Beta–Pascal	Below negative binomial	Above negative binomial

Given a set of data, we obviously would not expect exact plots of the form (5.38), but provided we use only those u_r for which, say $f_{r-1} \geqslant 5$, the plot will give a fair indication of the appropriate type of distribution. Whether or not the indicated model is reasonable must, as always, be decided by the investigator.

Successive u_r are clearly dependent, since replacing f_{r-1} by $f_{r-1} + 1$ raises u_{r-1} but lowers u_r. This suggests having recourse to a simple smoothing operation such as

$$v_r = \tfrac{1}{2}(u_r + u_{r+1}). \tag{5.41}$$

Following Gart (1970) we can evaluate the asymptotic variances for the u_r and v_r to explore the variability of the plots. If n observations are taken, (n_r/n) denotes the proportion corresponding to variate value r, such that

$$E\,(n_r/n) = f_r,$$

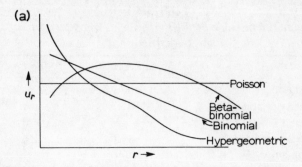

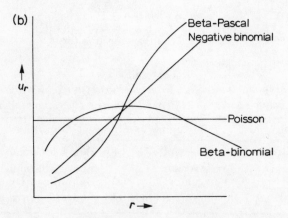

Fig.5.4   Plots of u_r ratios for some discrete distributions
(Adapted from Ord, 1967c)

$$\text{var}\,(n_r/n) \;=\; f_r(1 - f_r)/n\,,$$

and
$$\text{cov}\,(n_r/n,\, n_s/n) \;=\; -f_r f_s/n\,.$$

For the class of distributions under consideration, the estimators
$\tilde{u}_r = m_r/n_{r-1}$ converge in probability to u_r as $n \to \infty$, so that the usual
Taylor series argument yields, to $O(n^{-1})$,

$$\text{var}\,(\tilde{u}_r) \;=\; \frac{u_r^2}{n}\left\{\frac{1}{f_{r-1}} + \frac{1}{f_r}\right\},$$

$$\text{cov}\,(\tilde{u}_r,\, \tilde{u}_s) \;=\; -u_r u_{r+1}/n f_r\,, \quad s = r+1\,,$$
$$=\; 0\,, \quad s \neq r-1,\, r,\, r+1\,.$$

Similarly, it follows that if $\tilde{v}_r = \frac{1}{2}(\tilde{u}_r + \tilde{u}_{r+1})$,

$$\text{var }(\tilde{v}_r) = \frac{1}{4n}\left\{ u_{r+1}^2 \cdot \left(\frac{1}{f_r} + \frac{1}{f_{r+1}}\right) + u_r^2 \left(\frac{1}{f_{r-1}} - \frac{1}{f_r}\right)\right\},$$

and

$$\text{cov }(\tilde{v}_r, \tilde{v}_s) = \frac{1}{4n}\left\{\frac{u_r}{f_r}(u_{r+1} - u_r) - \frac{u_{r+1}}{f_{r+1}}(u_{r+2} - u_{r+1})\right\}, \quad s = r + 1,$$

$$= -\frac{u_{r+1}\,u_{r+2}}{4n\,f_{r+1}}, \quad s = r + 2,$$

$$= 0, \quad |r - s| > 2.$$

In particular, for the Poisson, all the u_r are equal, to u, say, and

$$\text{var }(\tilde{v}_r) = \frac{u^2}{4n}\left(\frac{1}{f_{r-1}} + \frac{1}{f_{r+1}}\right),$$

$$\text{cov }(\tilde{v}_r, \tilde{v}_{r+1}) = 0, \quad \text{cov }(\tilde{v}_r, \tilde{v}_{r+2}) = -u^2/4nf_{r+1}.$$

Having used the technique to select a distribution, could it be used for estimation? For (5.39) one might fit a straight line to selected u_r (or v_r) values by least squares, though such an approach is clearly rather arbitrary. Alternatively, one could select u_r values based on high observed f_r and solve sets of equations like (5.38) or (5.40). In both cases, the equations may be made linear in the unknown parameters. A more sophisticated method is given by McGilchrist (1969), who chooses optimal weights for linear functions of probabilities. Unfortunately, these weights are usually functions of the unknown parameter, so that the solution must be obtained iteratively. An alternative, non-iterative, approach would be along the lines of section 6.11.

Example 5.4

The table (drawn from Ehrenberg, 1959) shows the number of units of an item purchased during a 26-week period by members of a sample of 2000 consumers.

No. of units	0	1	2	3	4	5	6	7	8
No. of purchasers	1612	164	71	47	28	17	12	12	5

No. of units	9	10	11	12	13	14 to 26
No. of purchasers	7	6	3	3	5	8

The values of u_r and v_r are readily found using (5.38) and (5.41); we exclude the non-purchasers for the LSD model.

106

$$
\begin{array}{c}
r \\
u_r \\
v_r
\end{array}
\begin{bmatrix} 1 \\ 0\cdot10 \\ 0\cdot48 \end{bmatrix}
\begin{array}{cccccc}
2 & 3 & 4 & 5 & 6 & 7 \\
0\cdot87 & 1\cdot77 & 2\cdot67 & 3\cdot04 & 4\cdot24 & 7\cdot0\ldots \\
1\cdot32 & 2\cdot22 & 2\cdot86 & 3\cdot64 & 5\cdot62 & \ldots
\end{array}
$$

These are plotted in Fig. 5.5.

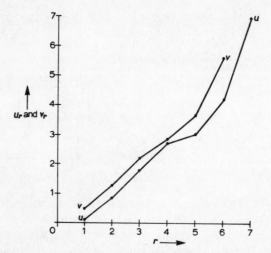

Fig.5.5 Plots of u_r and v_r for consumer purchases data

The plotted u_r and v_r ($r \geqslant 2$) strongly suggest a logarithmic series (LSD) or (truncated) negative binomial distribution (NBD); Ehrenberg (1959) presents these data in support of his LSD/NBD theory of consumer purchasing.

To estimate the parameter q for the LSD we could fit a straight line to the u_r and v_r (by least squares). We obtain $q = 0\cdot84$ using u_r ($r = 2,\ldots,6$) and v_r ($r = 2,\ldots,5$). Alternatively, we could use $u_2 = 0\cdot866$. The ML estimate is $\hat{q} = 0\cdot870$.

5.11 Truncated distributions

It is evident from (5.1) that the range of the random variable, R, may be truncated at either end, and the resulting truncated distribution will still be described by the differential equation. Likewise, the results for the plotting method in Table 5.4 are unchanged.

From section 5.6, Property D5, we observe that the moments of the truncated distributions depend upon both the unknown parameters and the terminal frequency(ies), and the distribution could, therefore, be fitted using the observed frequency(ies) and the sample moments; equations (5.11) are linear in the unknown parameters in this case and may readily be solved.

For example, when the range is $[u, \infty)$ and $b_0 = 0$, we obtain, using the observed frequency $\bar{f}_u$,

$$b_2^* \{(j + 2)m_{j+1} + 2j(j + 1)m_j + j^2(j - 1)m_{j-1} + u^{(2)} u^{(j)} \bar{f}_u\} +$$
$$+ b_1^* \{(j + 1)m_j + j^2 m_{j-1} + u \cdot u^{(j)}\bar{f}_u\} + a^*(m_j + jm_{j-1})$$
$$= m_{j+1} + (2j + 1)m_j + j^2 m_{j-1}, \quad j = 0, 1, 2; \quad (5.42)$$

where the m_j are the non-central factorial moments and $m_{-1} = 0$, $m_0 = 1$. a^*, b_1^*, b_2^* are the moment estimators.

Example 5.5

For the NBD, truncated at $u = 1$, we have

$$f_r = \binom{k + r - 1}{r} p^k q^r / (1 - p^k), \quad r = 1, \ldots$$

where $a = (k - 1)q/p$ and $b_1 = 1/p$. Then, from (5.42),

$$b_1^* = (m_2 + m_1 - m_1^2)/m_1(1 - \bar{f}_1)$$

and

$$a^* = m_1 + 1 - b_1^*(1 + \bar{f}_1),$$

so that the moment estimators for the parameters are

$$p^* = m_1(1 - \bar{f}_1)/s^2 \quad \text{and}$$
$$k^* = (m_1 p^* - \bar{f}_1)/(1 - p^*),$$

where s^2 is the sample variance.

For general $u \geq 1$, we find

$$p^* = \{m_1 - u\bar{f}_u(m_1 - u)\}/s^2 \quad \text{and}$$
$$k^* = (m_1 p^* - u\bar{f}_u)/(1 - p^*).$$

The estimators for $u = 1$ were discussed by Brass (1958) who showed that this approach was generally more efficient than using m_1, m_2 along with the analytic expression

$$f_1 = kp^k q/(1 - p^k),$$

which yields the pair of equations (Sampford, 1955)

$$m_2 p = m_1 q(k + 1) \quad \text{and}$$
$$m_1^2 p^{k+1} + s^2 p - m_1 = 0,$$

to be solved iteratively for p^*, k^*.

For k not too large, these estimators are reasonably efficient.

When $k \to 0$, the limit factor gives the logarithmic series distribution

$$f_r = \alpha q^r / r, \quad r = 1, 2, \dots, \tag{5.43}$$

where $\alpha^{-1} = -\ln(1 - q)$. The ML estimator $\hat{q}$ is the solution of $\bar{r} = \alpha\hat{q}(1 - \hat{q})^{-1}$, since the distribution is a GPSD (see section 6.4). Tables for $\hat{q}$, given $\bar{r}$, appear in Patil and Wani (1965), where the MLE is also tabulated for the range upper-truncated at d. Bowman and Shenton (1970) consider the distribution of $\hat{q}$ for small samples, showing a rapid approach to normality unless $q > 0 \cdot 9$.

The truncated Poisson is discussed in detail in Haight (1967, pp. 87–91) – the moment estimators follow from (5.42); see also David and Johnson (1952), Plackett (1953), and Cohen (1954; 1960a, b, c, d; 1961) and other articles referred to by Haight.

A general iterative technique for estimating the parameters of censored and truncated distributions is given by Hartley (1958), though the basic idea dates back to McKendrick (1926). The steps are as follows:

(i) obtain initial estimates of the parameters, e.g. by moments;

(ii) substitute these values into the density function to estimate the missing relative frequencies (allocating in proportion);

(iii) using these estimates for the missing densities, estimate the parameters by maximum likelihood (or other appropriate efficient method);

(iv) continue to iterate between (ii) and (iii) until the solution converges (which will happen, under very mild conditions). In practice, convergence is rapid when the "missing" frequencies account for less than about 30 per cent of the whole (see Exercise 5.17).

The inverse moments, $E(R^{-j})$, for truncated distributions, and their use as estimators have been analysed and tabulated by several authors, notably Grab and Savage (1954, binomial and Poisson); Tiku (1964, Poisson), Rider (1962, NBD), and Govindarajulu (1962, NBD; 1963, binomial; 1964, hypergeometric) – see Exercise 5.15.

5.12 Extensions of the difference equation system

The difference equation (5.1) has numerator of degree $j = 1$, denominator of degree $k = 2$. Davies (1934) considered the hypergeometric distributions generated by the terms of

$$f_0 \, F(\alpha, \beta; \gamma; \lambda),$$

corresponding to $j = k = 2$. More generally, this extension allows bimodal densities, generated by $f_0 \, F(\alpha, \beta, \epsilon; \gamma, \delta; 1)$. Several special cases are of interest:

(i) The confluent hypergeometric, $f_0 \, \phi(\alpha; \gamma; \lambda)$; this further reduces to the hyper-Poisson if $\alpha = 1$; see section 6.4.

(ii) The distribution of the number of people served during the "busy period" of a queueing process, when arrivals occur at random, is

(a) when service time is constant, the Borel–Tanner distribution (Tanner, 1953), i.e.

$$f_r = e^{-nq} \frac{(nq)^{n-r}}{(n-r)!} \frac{r}{n}, \quad r = 1, 2, \dots ; \qquad (5.44)$$

corresponding to the terms of $\phi(2; 1 - n; (nq)^{-1})$.

(b) When the service time is exponentially distributed, the Haight distribution (Haight, 1961), i.e.

$$f_r = \binom{2n - r - 1}{n - 1} r a^{n-r} / n (1 + a)^{2n-r}, \quad r = 1, \dots , \qquad (5.45)$$

corresponding to $F(2, 1 - n; 2 - 2n; 1 + \alpha^{-1})$.

When R takes on negative as well as positive values, this could arise as the difference between two non-negative random variables. Such distributions are often sharply peaked, but the Type IV (or VII) representation is unsuitable. Instead we may use

$$\Delta f_{r-1} = \frac{(a_0 - a_1 r) f_{r-1}}{b_0 + b_1 r + b_2 r^{(2)}} , \qquad (5.46)$$

where $a_0 = (1 - c) + c\delta(r),$

$$\delta(r) = 1, \text{ if } r \geq a, \quad = 0, \text{ otherwise.} \qquad (5.47)$$

This representation also allows the distribution to be bimodal. The two-sided geometric distribution, considered in Exercise 5.18, obeys (5.45).

5.13 Series expansions

The most important expansion is the discrete Cornish–Fisher

series obtained by Esseen (1945), for a lattice variable of width h. Further terms are given in Mayne (1972).

Following the notation of section 2.8, let R have cumulants $\{\kappa_j\}$, and let X be a normal variate, $N(\eta_1, \sigma^2)$ taking values x.

Put
$$\kappa_j - \eta_j = l_j\sigma^j, \quad \eta_2 = \sigma^2, \quad \eta_j = 0, \quad j \geqslant 3$$

and
$$\delta = h/c;$$

then $\quad l_1$ is $O(n^{-\frac{1}{2}})$, l_2 is $O(n^{-1})$ and $l_j \sim O(n^{1-\frac{1}{2}j})$, $\quad j \geqslant 3$,

where n is a parameter, generally sample size. Finally, put $\epsilon = l_1 - \frac{1}{2}\delta$, $\rho = l_2 - \frac{1}{12}\delta^2$; then x may be expressed in terms of $y = (r - \eta_1)/\sigma$ as

$$x = y - \sum_{s=1}^{m} X_s(y) + o(n^{-m/2}) \tag{5.48}$$

where $\quad X_1(y) = \dfrac{1}{6}l_3(y^2 - 1) + \epsilon$

$$X_2(y) = \frac{1}{72}\{3l_4(y^3 - 3y) - 2l_3^2(4y^3 - 7y) - 24l_3 y\epsilon + 36y\rho\}$$

$$X_3(y) = \frac{1}{6480}\{54l_5(y^4 - 6y^2 + 3) - 45l_4 l_3(11y^4 - 42y^2 + 15) +$$

$$+ 10l_3^2(69y^4 - 187y^2 + 52) + 810(2l_2 l_3 - l_4\epsilon)(y^2 - 1) +$$

$$+ 10l_3^2(216y^2\epsilon - 112\epsilon) - 540l_3(5y^2\rho - 3\rho - 2\epsilon^2) +$$

$$+ 1620l_2\delta - 3240\rho\epsilon\}.$$

The inverse series is

$$y = x - \sum_{s=1}^{m} Y_s(x) + o(n^{-m/2}) \tag{5.49}$$

where $\quad Y_1(x) = \dfrac{1}{6}l_3(x^2 - 1) + \epsilon$

$$Y_2(x) = \frac{1}{72}\{3l_4(x^3 - 3x) - 2l_3^2(2x^3 - 5x) + 36x\rho\}$$

$$Y_3(x) = \frac{1}{3240}\{27l_3(x^4 - 6x^2 + 3) - 135l_4 l_2(x^4 - 5x^2 + 3) +$$

$$+ 10l_3^2(12x^4 - 53x^2 + 17) +$$

$$+ 135l_3(6l_2 - 4\rho)(x^2 - 1) - 810l_2\delta\}.$$

Taking $l_1 = \frac{1}{2}\delta$, $l_2 = \frac{1}{12}\delta^2$, which yields $\epsilon = \rho = 0$, simplifies these expressions considerably, though these values are not necessarily the best ones to use.

Alternatively, the discrete orthogonal polynomials may be used (Hildebrandt, 1931; Beale, 1941; see Table A2 in Appendix A). If $p_k(r)$ denotes the kth orthogonal polynomial for density f_r, an arbitrary density g_r may be expressed in terms of f_r as

$$g_r = f_r\{1 + \sum_{k=1}^{\infty} a_k p_k(r)\};\qquad (5.50)$$

as usual, we shall be interested in series truncated after the first few terms. Based on the Poisson, with parameter λ, the Charlier Type B series (Charlier, 1905; Aroian, 1937; Kullback, 1947) is of this form with

$$p_k(r) = \sum_{j=0}^{k} \binom{k}{j} \lambda^{k-j}(-1)^j r^{(j)} \qquad (5.51)$$

and

$$k!\, a_k = \sum_{j=0}^{k} \binom{k}{j}(-1)^j \lambda^{k-j} \mu'_{(j)}, \qquad (5.52)$$

$\mu'_{(j)}$ being the jth factorial moment about the origin. The best choice of λ is not necessarily the mean of the g population (Boas, 1949). The Charlier polynomials have been generalised by Truesdell (1947).

Greenleaf (1932) used Krawtchouk polynomials for the symmetric binomial; this discussion has been extended by Gonin (1961) to cover all binomial and negative binomial distributions.

A similar approach is that of Wicksell (1934), who considered

$$g_r = \sum_{k=0}^{\infty} a_k \Delta^k f_{r-k}, \qquad (5.53)$$

which coincides with (5.51) when the f-distribution is Poisson. A special case of (5.52), when the f-distribution is the symmetric binomial, was considered by Thiele (1937).

EXERCISES

5.1 By defining the discrete distributions on a lattice of width h, show that as $h \to 0$ the Type III(d) distributions approach the gamma (Type III(c)), with scale parameters given by l. Show also that the discrete Types I(b) and VI approach their continuous analogues.

5.2 Show that when $f_r = \binom{k+r-1}{r} p^k q^r$ and we have prior information on p of the form

$$f(p) = p^{a-1}(1-p)^{b-1}/\beta(a, b)$$

the posterior distribution for R is of Type VI(d). Consider the case $a = b = 1$.

5.3   If there are n units in a closed system, and at time t, r of these n are "on" and $n - r$ are "off" with transition probabilities

$$p_r(+1) = \text{prob }\{(r + 1) \text{ on at } (t + dt) \,|\, r \text{ on at } t\}$$

$$= (n - r)pdt + o(dt)$$

$$p_r(-1) = \text{prob }\{(r - 1) \text{ on at } (t + dt) \,|\, r \text{ on at } t\}$$

$$= rq\, dt + o(dt); \quad r = 0, 1, \dots, n,$$

show that, when the system attains equilibrium, the probability that r units are on is given by the binomial distribution. Given the above system, but now assuming that

$$p_r(+1) = (r + 1)qdt + o(dt)$$

$$p_r(-1) = (n + 1 - r)pdt + o(dt)$$

show that the density function for r is of Type III(U), displaying the "bandwagon" effect.

5.4   For the Type VII(d), $f = a_k \Big/ \prod\limits_{j=r}^{k+r} \{(j + a)^2 + b^2\}$, denote the pth central moment by $\mu(p, k)$.
By considering the expression

$$\alpha_{k+2}^{-1} \cdot \mu(2p + 3, k + 2) - \alpha_k^{-1} \cdot \mu(2p - 1, k) -$$

$$- \tfrac{1}{2}\alpha_{k+2}^{-1}\{(m + 2)^2 - 4b^2\} \cdot \mu(2p + 1, k + 2),$$

show that, provided it exists, $\mu(2p + 3, k + 2) = 0$.
Further, show that unless $a = j/2$, j an integer, the density function is asymmetric.   (Ord, 1968b)

5.5   For the Polya urn scheme (5.7) define the random variable $x_i = +1$ if the ith ball drawn is black, $x_i = -1$ if the ith ball drawn is white. For the symmetric case $2M = N$, show that

$$\rho = \text{corr}(x_i, x_j) = c/(2M + c), \quad \text{for } i \neq j,$$

so that, if $\eta = (1 - \rho)/2\rho$,

$$f_r = \beta(r + \eta, \eta)/\beta(\eta, \eta). \quad \text{(McFadden, 1955)}$$

5.6   Use (5.1) to show that the probability generating function, $G(z)$, of any distribution in the difference equation system satisfies

$$(1 - z)b_2 z^2\, G''(z) + z\{b_1(1 - z) - 2zb_2 + z\}G'(z) +$$

$$+ \{b_0(1 - z) - zb_1 + (1 - a)z\}G(z) - E = 0,$$

where $\quad E = z^u f_u(b_0 + b_1 u + b_2 u^{(2)}) - z^{v+1}f_v\{b_0 + a + (v + 1)$
$$\times\ (b_1 - 1 + vb_2)\}.$$

5.7   Show that the β_1, β_2 coefficients for the Type III(d) distributions satisfy $\beta_2 - \beta_1 - 3 + 2y = 0$, where $y = n^{-1}, -k^{-1}, 0$ according as the distribution is binomial, negative binomial or Poisson. Further show that, for all three, $2\beta_2 - 3\beta_1 - 6 + 1/\mu_2 = 0$.

(E.S. Pearson, 1925; Merrell, 1933)

5.8   Show that, for a given mean value, the indices of dispersion of the discrete distributions obey the ordering $[I(x) \equiv I$ for Type $x]$

$$I[\text{I(a)}] < I[\text{III(B)}] < \begin{Bmatrix} I[\text{I(b)}] \\ I[\text{III(P)}] = 1 \end{Bmatrix} < I[\text{III(NB)}] < I[\text{VI}].$$

5.9   Use (5.19) to find the mean deviation for the hyper-Poisson family.

5.10   The binomial distribution has generating function $G(z) = (q + pz)^n$. Use (5.22) to derive the mean difference given in Table 5.2.

5.11   For the negative binomial with unknown parameters k and $m\ (\equiv kq/p)$, show that the ML estimators are the solutions of the equations

$$nm = \sum_{r=0}^{\infty} rn_r = n\bar{r}$$

$$n \ln (1 + \bar{r}/k) = \sum_{r=1}^{\infty} n_r \{1/k + 1/(k + 1) + \ldots + 1/(k + r - 1)\},$$

where n is the total sample size and n_r the number of observations with variate value r.

Show that $\hat{m}$ and $\hat{k}$ are asymptotically uncorrelated and find their asymptotic variances. (Anscombe, 1950)

5.12   The symmetric discrete Cauchy has $f_r = \alpha(r^2 + b^2)^{-1}$. Show that $\alpha = \pi b^{-1} \coth \pi b$, and the frequency moments, ω_i, are

$$\omega_2 = (2b^3)^{-1}\, \alpha^2 \pi\, [\coth \pi b + \pi b\ \text{cosech}^2\ \pi b],$$

$$\omega_3 = (8b^5)^{-1}\, \alpha^2 \pi\, [3 \coth \pi b + \pi b\ \text{cosech}^2\ \pi b\ \{3 + 2\pi b \coth \pi b\}].$$

Use these results together with those of section 1.10 to establish that ω_2 is asymptotically fully efficient for b. (Ord, 1968b)

114

5.13 Plot u_r from (5.38) to test the adequacy of a binomial model for the data below:

No. of trumps (r) in the first hand at whist (K. Pearson, 1924a)

r	0	1	2	3	4	5	6	7	8	9 or more	Total
Frequency	35	290	696	937	851	444	115	21	11	0	3400

(Ord, 1967c)

5.14 Estimating the NBD and LSD parameters from the data of Example 5.4 (using Fig. 5.4), compare the fit to the data of the NBD, LSD and NBD truncated at zero models, using u_r and v_r plots.

5.15 For the truncated NBD,

$$f_r = \binom{k + r - 1}{r} p^k q^r / (1 - p^k); \quad r = 1, 2, \ldots,$$

show that $E_{k,j} = E(R^{-j} \mid \text{parameter } k)$ satisfies

$$E_{k,j} = \frac{p(1 - p^{k-1})}{(k - 1)(1 - p^k)} \{(k - 1)E_{k-1,j} + E_{k-1,j-1}\}$$

and $E_{k,1} = (p^{-k} - 1)^{-1} \left\{ -\ln p + \sum_{j=1}^{k-1} \binom{k - 1}{j} (p/q)^j / j \right\}.$

(Rider, 1962)

5.16 Use the method of (5.42) to derive moment estimators for the Poisson distribution truncated (a) above, and (b) below, some value $r = N$.
(Rider, 1953; Cohen, 1961; Patil, 1962a; see also Haight, 1967, pp. 90–1).

5.17 Apply the iterative procedure of section 5.12 to fit a Poisson distribution to the following data on the number of major industrial stoppages (strikes) in the U.K. in the period 1948–1959 (i.e. to allow for the censoring of the upper tail).

No. of outbreaks	0	1	2	3 or more	Total
Observed no. of weeks	252	229	109	36	626

Repeat the procedure assuming that three outbreaks were observed in 28 weeks, but that the number of weeks without strikes is unknown (i.e. to allow for truncation of the lower tail).

(Richardson, 1944; M.G.Kendall, 1961)

5.18 Show that the two-sided geometric distribution defined by

$$f_r = c\rho^r, \quad r \geqslant 0$$
$$= c\nu^r, \quad r < 0$$

where $c = (1 - \rho)(1 - \nu)(1 - \nu\rho)^{-1}$ has mean $(\rho - \nu)/(1 - \rho)$ $(1 - \nu)$ and variance $\rho(1 - \rho)^{-2} + \nu(1 - \nu)^{-2}$. Further show that, when $\rho = \nu$,

$$\beta_2 = 4 + \frac{(1 + \rho)^2}{2\rho}.$$

(Ord, 1967a)

5.19 Given any density function f_r, $r = 0, 1, \ldots$, prove that

$$h_r = \frac{1}{1 - f_0} \sum_{j > r} f_j/j, \quad r = 0, 1, \ldots$$

is also a density function, with probability generating function

$$G_h(z) = \frac{1}{(1 - z)(1 - f_0)} \int_z^1 \left\{ \frac{G_f(u) - p_0}{u} \right\} du.$$

Hence find the mean and variance for the h-population in terms of the f-population. (The h-density is known as a STER density — **S**ums successively **T**runcated from the **E**xpectation of the **R**eciprocal.) (Bissinger, 1965; Patil and Joshi, 1968)

5.20 Find the STER distribution corresponding to the geometric distribution with density function $(1 - q)q^r$, $r = 0, 1, \ldots$. Conversely, show that the f-population corresponding to the geometric as a STER population has pgf

$$G_f(z) = f_0 + z(1 - f_0) \frac{(1 - q)^2}{(1 - qz)^2}.$$

(Kemp and Kemp, 1969)

CHAPTER 6

THE POWER SERIES AND CONTAGIOUS DISTRIBUTIONS

6.1 Introduction

Among continuous distributions the single parameter exponential family (see section 1.12) has the defining feature that the sample mean is the sufficient statistic. The discrete analogue of this family, the generalised power series distributions (GPSD), have the same property. We may write the density function as

$$\text{prob } (R = r) \equiv f_r \equiv f(r, \theta) = a(r)\theta^r/A(\theta), \qquad (6.1)$$

where $r \in \Omega$, the range of random variable R, and θ is a parameter. If there are other parameters, ϕ, we use the general form

$$f_r = a(r, \phi)\theta^r/A(\theta, \phi). \qquad (6.2)$$

As mentioned in section 5.1, the reduced difference equation system defined by

$$\Delta f_{r-1} = (a - r)f_{r-1}/rb_1 \qquad (6.3)$$

and the hyper-Poisson family (see section 6.4) both satisfy (6.2).

The other set of distributions discussed in this chapter are the so-called "contagious" distributions. As noted in section 4.1, these may correspond to either true or apparent contagion, and discrimination between the two models is not always possible. This difficulty may be resolved, however, by taking observations over more than one time-period — see section 7.8.

Many of the distributions considered in this chapter involve Poisson mixtures, and Haight (1967, Chapter 3) should be consulted in this connection. The dictionary and bibliography of Patil and Joshi (1968) is another valuable source of information. A brief review is also given by Kemp (1967b).

6.2 The generalised power series distributions (GPSD)

This class of distributions was first considered by Tweedie (1947), in the general form

$$f_r = a(r) \, \xi(r, \theta)/A(\theta), \quad r \in \Omega; \qquad (6.4)$$

116

taking $\quad \xi(r, \theta) = e^{-r\theta}, \quad r$ continuous,

$\qquad\qquad = \theta^r, \quad r$ discrete.

Tweedie calls these distributions Laplacian as $\int_\Omega e^{-r\theta} a(r)dr = 1/A(\theta)$

for r continuous, so that $\{A(\theta)\}^{-1}$ is the Laplace transform of $a(r)$. We list the properties of the GPSD's below, referring to the corresponding properties (E) of the exponential family where relevant (cf. section 1.12).

Property G1 (E1). Since $\sum_{r \in \Omega} a(r)\theta^r = 1/A(\theta)$, the probability generating function (pgf) is given by

$$G(z) = \Sigma f_r z^r = A(\theta z)/A(\theta). \qquad (6.5)$$

Property G2. From property E2, we see that the $(j + 1)$th cumulant, κ_{j+1}, obeys

$$\kappa_{j+1} = \frac{\theta d\kappa_j}{d\theta}, \quad j = 1, 2, \ldots;$$

further, from G1, we have for the factorial cumulants, $\kappa_{(j)}$, that (Khatri, 1959)

$$\kappa_{(j+1)} = \theta \frac{d\kappa_{(j)}}{d\theta} - j\kappa_{(j)}. \qquad (6.6)$$

Finally, for the non-central moments μ'_j, it may be shown (Noack, 1950) that

$$\mu'_{j+1} = \frac{\theta d\mu'_j}{d\theta} + \mu'_1\mu'_j. \qquad (6.7)$$

Property G3. From (6.4),

$$\frac{d \ln f_r}{d\theta} = \frac{r}{\theta} - \frac{A'(\theta)}{A(\theta)} = \frac{1}{\theta}(r - \kappa_1), \qquad (6.8)$$

since $\kappa_1 = \theta A'(\theta)/A(\theta)$, from G2.

Further, from (6.6),

$$\frac{d \ln f_r}{d\kappa_{(1)}} = \frac{(r - \kappa_{(1)})}{\theta} \cdot \frac{d\theta}{d\kappa_{(1)}} = \frac{r - \kappa_{(1)}}{\kappa_{(1)} + \kappa_{(2)}}. \qquad (6.9)$$

Corollaries (Patil and Shorrock, 1965)

(1) Within the exponential class, if $\kappa_1 = \kappa_2$, then the distribution is either Poisson (r discrete) or gamma (r continuous).

(2) If κ_1 is given, the distribution is uniquely determined within the class of GPSD's (cf. E4).

Property G4. For testing the null hypothesis $H_0 : \theta = \theta_0$ against a one-sided alternative, there exists a UMP test (see Property E3).

Property G5 (E5). A parametric function $\tau(\theta)$ is MVBU (minimum variance bound unbiased) estimable if and only if the derivative of the log-likelihood has the form

$$\frac{\partial \ln L(\theta)}{\partial \theta} = A(\theta) \{t(r) - \tau(\theta)\}. \qquad (6.10)$$

In consequence, for a GPSD, $\tau(\theta)$ is MVBU estimable if and only if $\tau(\theta) = a\kappa_1 + b$, a and b constants.

Several other properties are given in Patil (1962a).

6.3 Estimation for the GPSD

From (6.8) we see that the maximum likelihood (ML) estimator, $\hat{\theta}$, is the solution to

$$\bar{r} = \theta A'(\theta)/A(\theta),$$

which may be solved iteratively using the methods of Patil (1962b). For a sample of size n, the bias of $\hat{\theta}$ is, to $O(n^{-1})$,

$$b(\theta) = \frac{-\theta}{n} \{\mu_3(\theta) - \mu_2(\theta)\}/\mu_2(\theta),$$

where $\mu_j(\theta)$ is the jth moment of $\hat{\theta}$.

A uniform minimum variance unbiased (UMVU) estimator for θ^s was obtained by Roy and Mitra (1957). If

$$\left.\begin{array}{l} h_s(r) = 0, \quad \text{for } r < s, \\[2mm] \quad = \dfrac{a(r-s)}{a(r)}, \quad r \geqslant s, \end{array}\right\} \qquad (6.11)$$

it follows that $E\{h_s(r)\} = \theta^s$. Further, $T = \sum_{i=1}^{n} r_i$ is sufficient for θ since

$$f(\mathbf{r}, t|\theta) = f_1(t|\theta)f_2(\mathbf{r}|t), \quad \mathbf{r}' = (r_1, \ldots, r_n),$$

where

$$f_1(t|\theta) = \Pr(T = t) = c(t, n)\theta^t\{A(\theta)\}^{-n}, \qquad (6.12)$$

$$c(t, n) = \sum_{\Omega} \prod_{i=1}^{n} a(r_i), \qquad (6.13)$$

and $\Omega = \{r_i : r_i \text{ non-negative integers and } \sum_{i=1}^{n} r_i = t\}$.

If
$$u_s(t) = 0, \quad t < s,$$

$$= \frac{c(t - s, n)}{c(t, n)}, \quad t \geqslant s,$$

then it may be shown that $E\{u_s(t)\} = \theta^s$, thereby extending (6.11) to cover a sample of size n. As $u_1(t)$ is an unbiased estimator and also a function of the ML estimator, it follows that it is UMVU. Further, since $E(u_2) = \theta^2$, an unbiased estimator for the variance of u_1 is

$$\hat{\sigma}^2(u_1) = \{u_1(T)\}^2 - u_2(T). \tag{6.14}$$

Example 6.1 (Roy and Mitra, 1957)

For the negative binomial distribution (NBD),

$$f_r = \binom{k + r - 1}{r} p^k q^r$$

and the distribution of T is also a NBD with parameters kn, p, so that

$$c(t, n) = \binom{kn + t - 1}{t}.$$

Thus $u_1(T) = T/(kn + T - 1)$ and

$$\hat{\sigma}^2(u_1) = (kn - 1)T/\{kn + T - 1)^2(kn + T - 2)\}.$$

6.3.1 Estimation of the distribution function

The efficiency of $T = \Sigma \, r_i$ for θ for the GPSD's enables one to obtain a minimum variance unbiased (MVU) estimator for the distribution function. Because of sufficiency, the density function may be factored into

$$f(r_1, \ldots, r_n \,|\, \theta) = h(r_1, \ldots, r_n \,|\, t) \, g(t \,|\, \theta).$$

Using (6.12), $h(r_1, \ldots, r_n \,|\, t) \propto \{\prod_{i=1}^{n} a(r_i)\}/c(t, n)$, from which

$$h(r_i \,|\, t) = \frac{a(r_i) \, c(t - r_i, n)}{c(t, n)}. \tag{6.15}$$

If the distribution function is $F(a, \theta) \equiv \text{prob } (R \leqslant a)$, then its MVU estimator is

$$\tilde{F} = \sum_{r_i \leqslant a} h(r_i \,|\, t). \tag{6.16}$$

That $\tilde{F}$ is unbiased follows easily, since

$$E(\tilde{F}) = \sum_{r_i \leqslant a} \sum_t h(r_i \mid t) g(t \mid \theta)$$

$$= \sum_{r_i \leqslant a} f(r_i \mid \theta).$$

A similar result may be obtained for the exponential family.

6.3.2 Ratio and moments estimators

An alternative to the MVU estimators is the class of ratio estimators proposed by Patil (1962c). If n_r observations take the value r, put

$$n v_s(t) = \sum_{r=c+s}^{d} n_r \, a(r - s)/a(r), \qquad (6.17)$$

where the range of r is $[c, d]$. Unlike u_s, v_s may be biased, and when $s = 1$ this bias is

$$b(v_1) = \frac{\theta}{n} f_{d-1}(1 - f_{d-1})^{-2} + O(n^{-2}). \qquad (6.18)$$

The estimator has variance, to $O(n^{-1})$,

$$\mathrm{var}(v_1) = n^{-1}(1 - f_d)^{-2} \left\{ 2\theta^2 f_{d-1} - (1 - f_d)\theta^2 + \sum_{r=c+1}^{d} \left(\frac{a_{r-1}}{a_r}\right)^2 f_r \right\}, \qquad (6.19)$$

which reduces, when the estimator is unbiased, to the exact form

$$n^{-1} \left\{ \sum \left(\frac{a_{r-1}}{a_r}\right) f_r^2 - \theta^2 \right\}. \qquad (6.20)$$

An unbiased estimator for (6.20) is

$$\left\{ \sum \left(\frac{a_{r-1}}{a_r}\right) n_r^2 - n v_1^2 \right\} \Big/ n(n - 1).$$

A further possibility is to use the first two moments (Patil, 1962d). If

$$g_{ij} = \sum_{r=c}^{d-1} r^i \{(r + 1)a_{r+1}/a_r\}^j f_r,$$

then the first two moments of the distribution (6.1) are

$$\mu_1' \equiv \mu = \theta g_{01} + c f_c$$

and

$$\mu_2' = \mu + \theta g_{11} + c(c - 1) f_c.$$

Denoting the sample mean by m, and the jth sample moment about the origin by m_j, we obtain the estimator

$$\tilde{\theta} = \frac{m_2 - mc}{g_{11} - (c - 1) g_{01}}.\tag{6.21}$$

To $O(n^{-1})$, the bias is

$$b(\tilde{\theta}) = (\theta\sigma_{22} - \sigma_{12})/ng^2,\tag{6.22}$$

and the variance

$$\text{var}(\tilde{\theta}) = (\sigma_{11} - 2\theta\sigma_{12} + \theta^2\sigma_{22})/ng^2,\tag{6.23}$$

where $g = g_{11} - (c - 1)g_{01}$,

$$\sigma_{11} = (m_4 - m_2^2) + c^2(m_2 - m_1^2) - 2c(m_3 - m_1 m_2),$$

$$\sigma_{12} = (g_{31} - m_2 g_{11}) - c(g_{21} - m_1 g_{11}) -$$
$$- (c - 1)(g_{21} - m_2 g_{01}) + c(c - 1)(g_{11} - mg_{01}),$$

and $\sigma_{22} = (g_{22} - g_{11}^2) + (c - 1)^2(g_{02} - g_{01})^2 -$
$$- 2(c - 1)(g_{12} - g_{01}g_{11}).$$

When $c = 0$, these expressions reduce to

$$\tilde{\theta} = m_1/g_{01}, \quad g = g_{01},$$

$$\sigma_{11} = m_2 - m_1^2, \quad \sigma_{12} = g_{11} - m_1 g_{01} \quad \text{and} \quad \sigma_{22} = g_{02} - g_{01}^2.$$

Example 6.2

The truncated Poisson distribution has $f_r = \theta^r/(r!\,S)$, $S = \sum_{r=c}^{d} \theta^r/r!$. The left truncated distribution (d infinite) has the unbiased estimate for θ^s, from (6.17),

$$v_s = \sum_{r=c+s}^{\infty} r^{(s)} n_r;$$

and var$(v_1) = \{\theta + c(c + 1)f_{c+1}\}/n$ (cf. Plackett, 1953).

For the right truncated distribution ($c = 0$), (6.21) yields

$$\tilde{\theta} = \sum_{r=1}^{d} rn_r/(n - n_d),$$

with $$b(\tilde{\theta}) = \theta f_{d-1}/ng^2$$

and $$\text{var}(\tilde{\theta}) = \theta\{1 - f_d + \theta f_{d-1}\}/ng^2;$$

where $$g = 1 - f_d.$$

6.4   The hyper-Poisson family

We recall from section 5.3 that the hyper-Poisson distributions have p.d.f.

$$f_r = f_0\, \theta^r\, \frac{\Gamma(1 + b)}{\Gamma(1 + b + r)}, \quad r = 0, 1, \dots, \tag{6.24}$$

where $f_0 = 1/\phi(1; 1 + b; \theta)$ and ϕ is the confluent hypergeometric function. These distributions arise in certain birth-and-death processes (Hall, 1956). The hyper-Poisson class may be regarded as a special case of the class of confluent hypergeometric distributions defined by Bhattacharya (1966) with

$$f(r; \theta, \nu, 1 + b) = \frac{\Gamma(\nu + r)\, \Gamma(1 + b)}{\Gamma(1 + b + r)\, \Gamma(\nu)\, \phi(\nu; 1 + b; \theta)} \cdot \frac{\theta^r}{r!}. \tag{6.25}$$

Equation (6.25) reduces to (6.24) when $\nu = 1$, and is of interest in the theory of accident proneness.

When b is known, (6.24) defines a power series distribution and θ can be estimated by the methods of the previous section. Thus, we now consider the situation where both θ and b are unknown (θ known, b unknown is an unlikely combination, but may be dealt with using the mean, mean and zero frequency, or the first two moments).

The first three moments, which may be derived from (5.11), are

$$\left. \begin{aligned} \mu_1' \equiv \mu &= \theta - b(1 - f_0) \\ \mu_2' &= \theta(1 + \mu) - b\mu \\ \mu_3' &= \theta(1 + 2\mu) - \mu_2'(\mu_2' - \theta)/\mu. \end{aligned} \right\} \tag{6.26}$$

Various estimators have been proposed using the moments and the zero frequency (Bardwell and Crow, 1964, 1965; Staff, 1964, 1967). Letting m_j denote the jth non-central sample moment, we summarise these as follows. Using

(a) the first three moments:

$$\tilde{\theta}_1 = (m_3 m_1 - m_2^2)/(m_1 + 2m_1^2 - m_2)$$

$$\tilde{b}_1 = \{m_2 - \tilde{\theta}(1 + m_1)\}/m_1.$$

(b) the first two moments and the zero frequency:

$$\tilde{\theta}_2 = c\{(1 - \tilde{f}_0)m_2 - m_1^2\}$$

$$\tilde{b}_2 = c(m_2 - m_1 - m_1^2),$$

where $\quad c^{-1} = 1 - f_0(1 + m_1).$

(c) the first two moments:

$\tilde{\theta}_3$ is the solution of

$$\frac{(1 + m_1)\theta - m_2}{\theta - m_2 + m_1^2} = \phi\left\{1; \frac{(1 + m_1)\theta + m_1 - m_2}{m_1}; \theta\right\},$$

and $\tilde{b}_3$ is then found from the sample version of μ_2' in (6.26).

The ML estimators are the solutions to the equations

$$\phi(1; b + 1; \theta) = b/(b + m_1 - \theta)$$

and $\quad n \sum_{u=1}^{\infty} \frac{\theta^r \Gamma(1 + b)}{\Gamma(1 + b + u)} S(u) = \phi(1; 1 + b; \theta) \sum_{i=1}^{n} S(r_i),$

where $\quad S(u) = \sum_{i=1}^{u} (b + i)^{-1}.$

Empirical results in Bardwell and Crow (1965) suggest that the moments methods are nearly as effective as ML and are therefore to be preferred on computational grounds; see Table 6.5 (page 147) for an example. The asymptotic variances of $\tilde{\theta}_1$ and $\tilde{b}_1$ are given in the same paper.

6.5 Generalised Poisson distributions

Generalised distributions were defined in section 4.1. Following the notation of that section, if variate Y_i has density function $f_i(y)$, $i = 1, 2$, variate $X \sim Y_1 \vee Y_2$ has the generalised density $h(x) = f_1\{f_2(x)\}$. Thus, if Y_1 is a Poisson variate,

$$\text{prob } (X = x) = \sum_{s=0}^{\infty} e^{-\lambda}\frac{\lambda^s}{s!} \text{ prob } (Y_{21} + Y_{22} + \dots Y_{2s} = x). \qquad (6.27)$$

Then the pgf of X is $G(z) = \Sigma z^x \text{ prob } (X = x)$

$$= \sum_{s=0}^{\infty} e^{-\lambda}\frac{\lambda^s}{s!} \{g(z)\}^s,$$

where $g(z)$ is the pgf of Y_2, since prob $(Y_{21} + \dots + Y_{2s} = x)$ is the s-fold convolution of Y_2. Therefore

$$G(z) = \exp[\lambda\{g(z) - 1\}]. \qquad (6.28)$$

We refer to (6.28) as the pgf of the class of generalised, or sometimes "contagious", Poisson distributions (Feller, 1943). For example, if $g(z)$ is the pgf of the binomial $(q + pz)^m$, $G(z)$ is the pgf of the

124

Poisson ∨ binomial, or Poisson—binomial, distribution (see Table 6.1, pages 126—7).

The labelling of generalised Poisson distributions as "true contagion" processes arises as follows. Suppose that "clusters" of objects are observed, such as plants in a field or houses in a study area. Then the models proposed envisage a random (Poisson) distribution of such clusters, each cluster containing one or more objects, the number within a cluster following the generaliser distribution. In so far as the existence of a cluster means that an object is "more likely" to have other similar objects nearby, we may say that these processes represent "true contagion". A variety of these models appear in the literature and the main ones are listed in Table 6.1.

6.6 Accident proneness and compound Poisson distributions

Another model of considerable importance is the compound Poisson. As before, we consider the number of observations to be initially generated by a Poisson process, but instead of each observation being a cluster, we now suppose that different "areas" have different mean intensities or levels λ. That is, we assume that λ is itself a random variable and that its distribution may be specified. Thus the final distribution of the random variable R is the Poisson compounded with some other distribution. The notation used is, again, that of section 4.1, so that, for example, Poisson $\wedge$ gamma $\sim$ negative binomial.

These models are also used in the study of accidents — some individuals may be more prone to accidents than others, so that their rate of sustaining accidents, λ per unit time, is higher than that of other individuals. This concept of accident proneness was first used by Yule and Greenwood (1926) and Newbold (1927). The initial distribution need not be Poisson, though this assumption is generally made. The assumptions underlying the Poisson accident proneness model may be summarised as follows:

(a) A given individual, in a given time-period, has a constant liability, λ, of having an accident. This value, λ, is unaffected by the occurrence of accidents in any previous time-period, or by accidents sustained by other individuals.

(b) The number of accidents sustained by the individual in the given time-period is Poisson distributed with mean λ.

(c) The liability, λ, varies between persons, but is constant for each person in the given time-period.

The subject of accident proneness is too large to be treated fully in this monograph; for reviews see Arbous and Kerrich (1951), Irwin (1964), and Kemp (1970). The compound Poisson model is sometimes referred to as displaying "apparent contagion", since the negative binomial may be so generated without there being any clustering of the observations.

The pgf of the compound Poisson distribution may be written as

$$\int_0^\infty e^{\lambda(z-1)} f(\lambda, \mu) \, d\lambda = G(z - 1, \mu), \qquad (6.29)$$

where μ is the vector of parameters of the compounding distribution f. From the uniqueness of the Laplace transform we see that for any G there is a corresponding f. Also if f is a density function, and the jth moment of λ is μ'_j, then the jth factorial moment of $Y_1 \wedge Y_2$ is $m'_{(j)} = \mu'_j$. On specifying some G, we see that it is the pgf of a compound Poisson if and only if the corresponding f is a properly defined density function. Further, if G is the pgf of a generalised Poisson distribution and f is indeed a density function, then G specifies both a generalised and a compound Poisson form. This implies that the true and apparent contagion models cannot be distinguished on the basis of observations over a single time-period, and we must take observations over two or more intervals of time (see section 7.8).

Example 6.3

The Pólya—Aeppli distribution has pgf (from Table 6.1)

$$G(z) = \exp [\mu \{(1 - \rho u)^{-1} - 1\}],$$

where $\rho = \tau / (1 - \tau)$, $u = z - 1$. This may be written $G(z) = e^{-\mu} \sum_{j=0}^\infty \mu^j (1 - \rho u)^{-j}/j!$. But $(1 - \rho u)^{-j}$ is the moment generating function of the gamma distribution with density function

$$h(x, j, \rho) \propto x^{j-1} e^{-x/\rho};$$

therefore $f(\lambda)$ is the density function of the gamma $\wedge$ Poisson, with Poisson compounding on the gamma parameter j. This yields

$$f(\lambda) = \exp (- \mu - \lambda/\rho) \sum_{j=1}^\infty \frac{\mu^j \lambda^{j-1}}{\rho^j j! \, \Gamma(j)}, \qquad \lambda > 0,$$

$$= e^{-\mu}, \quad \lambda = 0.$$

For $\lambda > 0$ we could write $f(\lambda)$ as $i \exp (- \mu - \lambda/\rho)(\mu/\lambda\rho)^{\frac{1}{2}} \times J_1 (2i\omega)$, where $i^2 = - 1$, $\omega^2 = \mu\lambda/\rho$ and J_1 is the Bessel function

Table 6.1   *Compound and generalised*

Name of distribution (plus references)	Compound or generalised form (P = Poisson)
Negative binomial (Pascal) (Fisher, 1941; Feller, 1943; Quenouille, 1949)	$P \vee$ logarithmic $P \wedge$ gamma
Pólya–Aeppli (Pólya, 1931)	$P \vee$ geometric Pascal $\underset{k}{\wedge}$ Poisson
Neyman Type A (two parameter) (Neyman, 1939; Feller, 1943)	$P \vee P$ $P \wedge P$
Thomas (Thomas, 1949)	$P \vee P$ (one "parent" plus random number of offspring)
Poisson–binomial (McGuire *et al.*, 1957, Sprott, 1958)	$P \vee$ binomial Binomial $\wedge P$
Poisson–Pascal (Katti and Gurland, 1961; Shumway and Gurland, 1960b)	$P \vee$ Pascal Pascal $\wedge P$
Hermite (Kemp and Kemp, 1965, 1966; Patil, 1964)	$P \vee$ binomial ($m = 2$) $P \wedge$ truncated normal Normal $\vee P$
Beall–Rescia (Beall and Rescia, 1953)	$P \vee P \wedge$ beta $(1, \beta)$ $P \wedge P \wedge$ beta $(1, \beta)$
Gurland (Gurland, 1958)	$P \vee P \wedge$ beta (α, β) $P \wedge P \wedge$ beta (α, β)
Neyman Type A (four parameters)	$P \vee$ (finite mixture of two Poissons)
Short (Cresswell and Froggart, 1963; Kemp, 1967a)	$P \vee$ (Poisson with zeros added)

Poisson distributions

Factorial cumulant generating function ($u = z - 1$)	Factorial cumulants
$-k \ln (p - qu)$	$k(j - 1)! \, (q/p)^j$
$\lambda (1 - \tau)/(1 - \tau - \tau u)$	$j! \lambda \{\tau/(1 - \tau)\}^j$
$\lambda_1 \exp (\lambda_2 u)$	$\lambda_1 \lambda_2^j$
$\lambda_1 (u + 1) \exp (\lambda_2 u)$	$\lambda_1 \lambda_2^{j-1}(j + \lambda_2)$
$\lambda (1 + pu)^m$	$\lambda m^{(j)} p^j$
$\lambda p^k (p - qu)^{-k}$	$\lambda (k + j - 1)^{(j)}(q/p)^j$
$\lambda_1 u + \lambda_2 u(u + 2)$	$\kappa_{(1)} = \lambda_1 + 2\lambda_2; \; \kappa_{(2)} = 2\lambda_2$ $\kappa_{(j)} = 0, \, j \geqslant 3$
$\lambda_1 \Gamma(\beta + 1) \sum\limits_{j=0}^{\infty} \lambda_2^{j+1} u^{j+1} / \Gamma(\beta + j + 2)$	$j! \, \lambda_1 \lambda_2^j /(j + \beta + 1)^{(j)}$
$\lambda_1 \phi(\alpha; \alpha + \beta; \lambda_2 u)$	$\lambda_1 \lambda_2^j (\alpha + j)^{(j)} /(\alpha + \beta + j)^{(j)}$
$\lambda_1 p \exp (\lambda_2 u) + \lambda_1 q \exp (\lambda_3 u)$	$\lambda_1 (p\lambda_2^j + q\lambda_3^j)$
$\lambda_1 (e^{u\mu} - 1) + \lambda_2 u$	$\kappa_{(1)} = \lambda_1 \mu + \lambda_2$ $\kappa_{(j)} = \lambda_1 \mu^j, \, j \geqslant 2$

of the first kind of order one. This result may be written as
Poisson $\wedge$ gamma $\wedge$ Poisson, which is simply Pascal $(k, p) \underset{k}{\wedge}$
Poisson.

This example may be generalised to cover the Poisson–Pascal,
where the density function is still the gamma $\wedge$ Poisson, but $G(z)$
contains terms of the form $(1 - \rho u)^{-kj}$. Further examples appear as
Exercises 6.5 and 6.6.

This link between compound and generalised distributions was
first explored by Feller (1943) and is neatly summarised by the
following

Theorem 6.1 (Gurland, 1957)

Let Y_1 be a random variable with pgf $g_1(z) = \{h(z)\}^\theta$, where θ is a
given parameter. Suppose now that θ is the random variable Y_2 with
function F_2 and pgf g_2. Then, whatever the form of F_2,

$$Y_1 \wedge Y_2 \sim Y_2 \vee Y_1.$$

Proof The pgf of $Y_1 \wedge Y_2$ is $\int_{-\infty}^{\infty} \{h(z)\}^{cx}\, dF_2(x)$, while the pgf of
$Y_2 \vee Y_1$ is

$$g_2\{g_1(z)\} = \int_{-\infty}^{\infty} \{h(z)\}^{\theta x}\, dF_2(x).$$

These pgf's exist, at least for $z = e^{iv}$, and are equal when $\theta = 0$,
proving the theorem.

This theorem yields directly

Poisson $\wedge$ Poisson $\sim$ Poisson $\vee$ Poisson $\sim$ Neyman Type A

Pascal $(k, p) \underset{k}{\wedge}$ Poisson $\sim$ Poisson $\vee$ Pascal $\sim$ Poisson–Pascal

Pascal $(k, p) \underset{k}{\wedge}$ gamma $\sim$ gamma $\vee$ Pascal.

An extension of these results is given as Exercise 6.7.

6.7 Distributions derived from the Poisson

The density functions of important distributions which arise from
compounding or generalising the Poisson are listed in Table 6.1,
together with their factorial cumulants and generating functions.

Many relationships exist among the distributions listed in Table
6.1, some of which are recorded here.

(1) Since the Poisson is a limiting form of the binomial and negative binomial distributions, the Neyman Type A is the limiting form of the Poisson—binomial and Poisson—Pascal.

(2) The negative binomial and the Pólya—Aeppli are special cases of the Poisson—Pascal for $k \to 0$ and $k = 1$ respectively. .

(3) The Poisson $\lor$ GPSD form contains most of the distributions in the table as special cases (see Exercise 6.4).

(4) The four-parameter Neyman Type A reduces to the three-parameter form when $p = \frac{1}{2}$, and to the two-parameter form when $p(1 - p) = 0$.

(5) The Neyman Types A, B and C are all contained in the Beall—Rescia set, for $\beta = 0, 1, 2$ respectively. The Beall—Rescia family, in turn, corresponds to the Gurland system with $\alpha = 1$.

(6) A special case of the four-parameter Neyman Type A is used by Katti and Sly (1965), with parameters λ_1, $\lambda_2 = \lambda\alpha$, $\lambda_3 = \lambda(1 - \alpha)/m$, $p = (m + 1)^{-1}$, m known. This model assumes that members of a cluster (larvae on a stalk, for example) leave their cluster with probability α and migrate to one of m symmetrically placed neighbouring clusters with probability $1/m$.

(7) In addition to the compound forms listed in Table 6.1, the Poisson $\land$ lognormal and the Poisson $\land$ uniform have also been used (see Exercises 6.8, 6.15).

6.7.1 Expansion of the pgf

The pgf of a generalised Poisson may be written as

$$G(z) = \exp [\lambda\{g(z) - 1\}]. \tag{6.30}$$

Expanding (6.30) in powers of z often yields an infinite series for individual probabilities. To combat this, we rewrite (6.30) as

$$G(z) = \exp\left[\lambda(1 - p_0)\left\{\frac{g(z) - p_0}{1 - p_0} - 1\right\}\right], \tag{6.31}$$

where $p_0 = g(0)$. Then

$$\frac{g(z) - p_0}{1 - p_0} = zg^*(z),$$

so that

$$G(z) = \exp\{-\lambda(1 - p_0)\} \sum_{r=0}^{\infty} \frac{z^r\{g^*(z)\}^r \{\lambda(1 - p_0)\}^r}{r!}$$

and a finite series for f_r can always be found.

Example 6.4

From Table 6.1, it is seen that the pgf of the Poisson–Pascal is

$$\exp\left[\lambda\{p^k(1 - qz)^{-k} - 1\}\right],$$

which may be rewritten as

$$\exp\left[\lambda(1 - p^k)\left\{\frac{p^k(1 - qz)^{-k} - p^k}{1 - p^k} - 1\right\}\right].$$

This can be factorised as suggested, and terms in z^r then collected to give f_r. This procedure yields

$$f_r = q^r \exp\{\lambda(p^k - 1)\} \underset{T}{\textstyle\sum}\left\{\prod_{j=1}^{r} \theta_j^{i_j}/i_j!\right\}$$

where

$$\theta_j = \lambda p^j \binom{k + j - 1}{j} \quad \text{and} \quad T = \{i_j, j = 1, \ldots, r \mid \textstyle\sum_{j=1}^{r} j i_j = r\}.$$

For example,

$$f_0 = \exp\{\lambda(p^k - 1)\}$$

$$f_1 = q\theta_1 f_0$$

$$f_2 = q^2 f_0 (\theta_2 + \tfrac{1}{2}\theta_1^2)$$

and

$$f_3 = q^3 f_0 (\theta_3 + \theta_1 \theta_2 + \tfrac{1}{6}\theta_1^3).$$

Taking $k = 1$ gives an expansion for the Pólya–Aeppli, while the Poisson limit to the Pascal leads to an expansion for the Neyman Type A (two-parameter case). Other expansions can be obtained in the same way.

A further point of interest arising from (6.31) is that the generalised Poisson resulting from generalising with $g(z)$ is indistinguishable from that obtained using the truncated $zg^*(z)$.

Alternatively, the probabilities may be evaluated using a recursive relation given by Gurland (1965). If $G(z) = \Sigma f_r z^r$ and $g(z) = \Sigma p_r z^r$, repeated differentiation of (6.30) yields

$$f_r = (\lambda/r) \sum_{j=0}^{r-1} (j + 1) p_{j+1} f_{r-1-j}. \tag{6.32}$$

6.7.2 *Evaluation of cumulants*

If the distribution with pgf $g(z)$ has non-central moments μ_j', $j = 1, 2, \ldots$, then from (6.30) repeated differentiation yields

$$\kappa_j = \lambda\mu_j', \quad j = 1, 2, \ldots; \tag{6.33}$$

with a similar relation holding for factorial moments and cumulants. Since $\mu'_j \geqslant \mu'_i$ for $j > i$, we see that $\kappa_j > \kappa_i$; further, equality holds only for the Poisson distribution itself.

6.7.3 Modality

Barton (1957) showed that the Neyman Type A could have several modes, or even infinitely many. On the other hand, the negative binomial is unimodal, and Anscombe (1950) conjectured that the Poisson $\wedge$ lognormal was also unimodal. These results have been drawn together by Holgate (1970) in the following

Theorem 6.2

Let $f(\lambda)$ be the density function of a positive continuous unimodal random variable. Then the compound Poisson with pgf

$$\int_0^\infty e^{\lambda(z-1)} f(\lambda) d\lambda \quad \text{is unimodal.}$$

When $f(\lambda)$ is multimodal, the resulting compound Poisson may also be multimodal. Further, a discrete $f(\lambda)$ may be regarded as an extreme case of a multimodal continuous form, leading to a multimodal compound Poisson, confirming Barton's result.

6.8 Other generalised and compound distributions

So far the discussion has been restricted to Poisson models; these are the forms most commonly used, but other systems are sometimes of value. In particular, following the argument of section 6.5 we see that the generalised GPSD derived from (6.4) has the pgf

$$G(z) = \frac{A\{\theta\, g(z)\}}{A(\theta)}, \tag{6.34}$$

$g(z)$ being the pgf of the generaliser. The factorial cumulants are

$$\kappa_{(1)} = \frac{\partial G}{\partial z}_{z=1} = \theta A' g'/A = k_{(1)} m_1$$

$$\kappa_{(2)} = k_{(1)} m_2 + m_1^2 k_{(2)} - k_{(1)}^2 m_1^2,$$

where the $k_{(i)}$ are the factorial cumulants of the GPSD, and the m_i are the non-central moments of the generaliser.

Khatri and Patel (1961) have suggested three general types of discrete family, with pgf's

$$
\left.
\begin{aligned}
G_A(z) &= \exp\{h(z) - 1\}, \\
G_B(z) &= \{h(z)\}^n, \\
G_C(z) &= c \log\{h(z)\}.
\end{aligned}
\right\}
\tag{6.35}
$$

and

Type A includes all the generalised Poisson forms considered so far, plus such generalisations as the Poisson–hypergeometric. The most important classes in Type B are the generalised binomial and generalised Pascal, while Type C includes the generalised logarithmic series distributions. Recurrence relations may be derived for these distributions along the lines of section 6.7.2, while the cumulants are given in Exercise 6.10.

Other compound distributions of interest are the binomial $\underset{p}{\wedge}$ beta $\sim$ beta–binomial and the Pascal $\underset{p}{\wedge}$ beta $\sim$ beta–Pascal, discussed in section 5.3.

6.9   Choosing a particular model

When fitting a distribution to data, we need to decide which model is appropriate; two approaches are possible:

(1) Fit several alternative models to data and select that one which is the "best fit", as judged by chi-square or other appropriate criterion.

(2) Test the data before fitting and select a model on this basis.

Both procedures raise problems. If (1) is used, what does the calculated value of χ^2 signify? It is the most probable of several observed χ^2 values, each evaluated under its own null hypothesis, and each correlated to an unknown extent with the others. If (2) is used, have we lost any degrees of freedom? A partial answer to the second question may be formulated as follows:

Suppose we test H_0 : distribution function is $F(x \mid \theta = 0)$ against $H_1 : F(x \mid \theta \neq 0)$, using a test statistic which is a function of the ML estimator, based on grouped data, for θ (given H_1). Then we may regard this, approximately, as causing the loss of one degree of freedom, as the estimator for θ

$$= \text{grouped data ML estimator } (H_0 \text{ rejected}),$$
$$= 0 \ (H_0 \text{ not rejected}).$$

Thus, when we come later to test the goodness of fit, we should deduct the number of degrees of freedom lost as though the more general $F(x \mid \theta)$ were fitted.

This approach is in contradistinction to the first approach, where the two models would be tested for goodness of fit using different numbers of degrees of freedom. In general, we are often using inefficient estimators and test criteria which are not functions of the ML estimator, so that even the limited argument above is not valid. However, the general direction of the effect of procedure (2) can at least be noted.

Frequently we wish to choose between distributions when one is not a special case of the other, but both may be embedded in a more general family, and the test is then a comparison of two simple hypotheses. For example, the negative binomial (NB) and Pólya–Aeppli (PA) are special cases of the Poisson–Pascal, and Anscombe (1950) has shown that the third sample moment is asymptotically equivalent to the likelihood ratio criterion for testing $H_0 : F$ is NB against $H_1 : F$ is PA, as $p \to 1$ with k fixed (using the notation of Table 6.1). This suggests the use of the first three sample moments as suitable statistics for a crude procedure of type (2) with respect to the class of generalised Poisson distributions. Ord (1970) used the ratios $I = \mu_2/\mu_1'$, $S = \mu_3/\mu_2$, also considered in sections 1.7 and 5.7. Under the null hypothesis that the distribution is Poisson, I is approximately distributed as χ^2 with $(n - 1)$ degrees of freedom when n observations are available. The power of the test against various compound Poisson alternatives is discussed by Darwin (1957).

The I, S region is shown in Fig. 6.1, and the theoretical relationships between I and S for the distributions are summarised in Table 6.2.

We note the following points from Fig. 6.1:

(1) The PB boundaries are the line $S = I$ and the Neyman Type A (NTA) curve.

(2) The PP boundaries are the line $S = 2I$ and the NTA curve.

(3) Between the lines $S = 2I$ and $S = 2I - 1$ (the NB line) only J-shaped distributions occur, for which the mean (μ) is less than 1.

(4) The curve for the Thomas distribution is not shown on the diagram as it is very close to the NTA curve, which it approaches asymptotically. This bears out the statement that the NTA can be used as an approximation to the Thomas distribution for large λ_2.

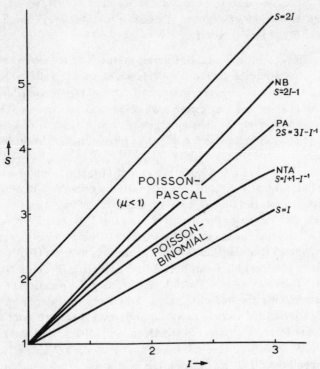

Fig.6.1   The I, S chart for the contagious discrete distributions
(Adapted from Ord, 1970)

Table 6.2

I, S relationships for generalised Poisson distributions, and values of η_j

Distribution	Code for Fig. 6.1	Relation (I, S)	η_j
Poisson	P	$S = I = 1$	1
Negative binomial	NB	$S = 2I - 1$	jw
Pólya–Aeppli	PA	$2S = 3I - I^{-1}$	$(j + 1)\tau/(1 - \tau)$
Neyman Type A	NTA	$S = I + 1 - I^{-1}$	λ_2
Poisson–Pascal	PP	$S = I + (1 - I^{-1})/(1 - q)$	$(k + j)w$
Poisson–binomial	PB	$S = I + (1 - I^{-1})(1 - p)$	$(m - j)p$
Hermite	H	As PB, $p = 2\lambda_2(\lambda_1 + 2\lambda_2)^{-1}$	
Thomas	T	$S = I + 1 - I^{-1}\left\{1 + \left(\dfrac{\lambda_2}{1 + \lambda_2}\right)^2\right\}$	$\lambda_2\{1 + (j - 1 + \lambda_2)^{-1}\}$

$$w = q/p$$

(5) For fixed l, we have the ordering $S(PP) > S(NTA) > S(PB)$.

(6) The Hermite (H) curve is restricted to $1 \leqslant l \leqslant S \leqslant 2$; the PB curves obey $1 \leqslant l \leqslant S \leqslant m$.

An alternative procedure, suggested by Hinz and Gurland (1967), is to consider the plot $\eta_j = \kappa_{(j+1)}/\kappa_{(j)}$ against j. The theoretical values of η_j are given in Table 6.2. The plot increases for NB and PP, is steady for NTA and falls for PB.

Having selected a particular model we must then estimate the parameters of that model, and we now consider some of the methods which can be employed.

6.10 Moments estimators

Using the results of Table 6.1 (pp. 126–7), estimators based on the first few moments and/or frequencies may be devised. In general, when the observed proportion of zeros (n_0/n) is high, estimators based on both moments and frequencies (such as n_0/n and the sample mean for two-parameter distributions) are highly efficient, while the all-moment estimators improve in accuracy as the sample mean moves away from the origin (see, for example, Anscombe's discussion of estimators for the negative binomial; Anscombe, 1950).

Many of the results about bias and efficiency are asymptotic to $O(n^{-1})$, but series expansions in decreasing order of n may be derived using orthogonal polynomials (see Shenton and Myers, 1965a, b; Shenton and Bowman, 1967).

6.11 Minimum chi-square (MCS) estimators

Maximum likelihood estimators may be derived in the usual way for all the distributions discussed in this chapter, but the resulting expressions are often extremely difficult to solve. An important exception to this general statement is considered in the next section. On the other hand, MCS estimators are asymptotically efficient and may be easier to apply. Gurland (1965) and Hinz and Gurland (1967) have developed modified MCS estimators for the class of generalised Poisson distributions considered in section 6.9. Their results are summarised below. These estimates have the advantage that they can be obtained non-iteratively, without sacrificing much in terms of asymptotic efficiency (small-sample properties await further investigation).

Put $T_j = f_j/f_0$ and define τ_j by

$$\log G(z) = \log f_0 + \sum_{j=1}^{\infty} \tau_j \, z^j/j! = \log f_0 + \log \sum_{j=0}^{\infty} T_j \, z^j. \quad (6.36)$$

Then
$$\tau_n = n!\, T_n - \sum_{j=1}^{n} (n-1)^{(j)}\, T_j \tau_{n-j}, \qquad (6.37)$$

and, in particular,

$$\tau_1 = T_1, \ \tau_2 = 2T_2 - T_1^2, \ \tau_3 = 6T_3 - 6T_2 T_1 + 2T_1^3,$$
$$\tau_4 = 24(T_4 - T_3 T_1 + T_2 T_1^2) - 12T_2^2 - 6T_1^4.$$

Further, put $\gamma_j = \tau_{j+1}/\tau_j, \quad j = 1, 2, \ldots$

and $\qquad \gamma_0 = \kappa_{(1)}.$ $\qquad\qquad (6.38)$

As in section 6.9, we also put $\eta_j = \kappa_{(j+1)}/\kappa_{(j)}$, $\eta_0 = \kappa_{(1)}$. The values of γ_j, $\kappa_{(j)}$ and η_j are given in Table 6.3 for the negative binomial, Neyman Type A, Poisson–Pascal and Poisson–binomial distributions, with the corresponding sample notation. The γ_j, like the η_j, can be plotted as an aid to selecting a particular model; the form of these plots is apparent from Table 6.3.

If s statistics are used to estimate b parameters $(\theta_1, \ldots, \theta_b)$, $s \geq b$, then the extended MCS method of Barankin and Gurland (1951) yields the linear relationship

$$\underset{(s \times 1)}{\xi} = \underset{(s \times b)}{W} \underset{(b \times 1)'}{\theta} \qquad (6.39)$$

where $\xi_j = \eta_j$ or γ_j for the distribution under consideration and W is known. Thus the MCS estimators of θ are in the form of generalised least-squares estimators

$$\theta^* = (W'\Lambda^{-1}W)^{-1}W'\Lambda^{-1}x, \qquad (6.40)$$

where x_j is the sample value of ξ_j and Λ is the covariance matrix of ξ. For practical purposes, Λ is replaced by a consistent estimator, Λ^* say.

Three alternative procedures are now listed, the choice depending on the particular distribution to be fitted. The θ should be chosen as functions of the original parameters so that the ξ are linear functions of them. For the negative binomial, for example, we would take $\theta_1 = kq/p$ and $\theta_2 = q/p$ (cf. Table 6.3). Because the estimators are of the form (6.40) their efficiency, as measured by the generalised variance [see equation (5.32)] is

$$E = |\tilde{I}|/|W'\Lambda W|,$$

where $\tilde{I}$ is the ML information matrix.

6.11.1 Using η_j only

Starting with the vector of non-central moments $M' = (\mu_1', \ldots, \mu_s')$,

Table 6.3

Value of coefficients for MCS estimators

| Population parameter: | $\kappa_{(j)}$ | η_j | γ_j | α |
Sample value:	$k_{(j)}$	h_j	l_j	a
Distribution				
Negative binomial	$(j-1)!\,kw^j$	jw	jq	$f_0(1+\eta_1)^{1+u} - 1$
Neyman Type A	$\lambda_1\lambda_2^j$	λ_2	$\lambda_1\lambda_2^j$	$f_0\eta_1\exp\{-u(e^{-\eta_1}-1)\}$
Poisson–Pascal	$\lambda w^j(k+j-1)^{(j)}$	$(k+j)w$	$(k+j)q$	$\left(1+\dfrac{\eta_1-v}{\eta_0}\log f_0\right)(v+1)^{\eta_1/v} - 1$
Poisson–binomial	$\lambda p^j m^{(j)}$	$(m-j)p$	$(m-j)p/q$	$f_0\eta_1\exp[u(1-m^{-1})\{1-\eta_1(m-1)^{-1}\}^m - 1\}]$

$$w = q/p; \quad u = \eta_0/\eta_1; \quad v = \eta_2 - \eta_1.$$

we wish to transform $\mathbf{M}$ to $\xi = \eta$. This can be done in three stages:

$$\mathbf{M} \rightarrow \kappa \text{ (vector of cumulants)}$$
$$\rightarrow \kappa^* \quad \text{(vector of factorial cumulants)}$$
$$\rightarrow \eta.$$

The Jacobians of these transformations may be written as the determinants of matrices $\mathbf{J}_1$, $\mathbf{J}_2$ and $\mathbf{J}_3$ respectively, where the matrices have elements

$$\mathbf{J}_1^{-1}(i, j) = \binom{i}{j}\mu'_{i-j}, \quad j \leqslant i,$$

$$= 0, \text{ otherwise,}$$

$$\mathbf{J}_2(i, j) = S_j^i, \quad j \leqslant i,$$

$$= 0, \text{ otherwise,}$$

where $\{S_m^n\}$ are the Stirling numbers of the first kind, listed for $m \leqslant n \leqslant 6$ in Table 6.4, and

Table 6.4
Stirling numbers of the first kind

n/m	1	2	3	4	5	6
1	1					
2	−1	1				
3	2	−3	1			
4	−6	11	−6	1		
5	24	−50	35	−10	1	
6	−120	274	−225	85	−15	1

$$\mathbf{J}_3(i, j) = 1/\kappa_{(i)} \quad \text{if } j = i + 1; \; \kappa_{(0)} = 1,$$

$$= -\kappa_{(i+1)}/\kappa_{(i)}^2, \quad 2 \leqslant j = i \leqslant s,$$

$$= 0, \text{ otherwise.}$$

Then $\text{var}(\eta) = \mathbf{J}\mathbf{V}\mathbf{J}'$, where $\mathbf{J} = \mathbf{J}_3\mathbf{J}_2\mathbf{J}_1$ and $\mathbf{V}$ is the covariance matrix of $\mathbf{M}$,

$$\mathbf{V}(i, j) = \frac{1}{n}\{\mu'_{i+j} - \mu'_i\mu'_j\}. \tag{6.41}$$

To estimate the $\mathbf{\theta}$, we replace the elements of $\mathbf{J}$, $\mathbf{V}$, $\mathbf{M}$ and $\mathbf{\eta}$ by consistent estimators based on $k_{(j)}$, h_j and l_j (see Table 6.3 for definitions).

6.11.2 Using γ_j only

The same approach may be adopted for $\mathbf{\xi} = \mathbf{\gamma}$, when

$$\Lambda = \Omega = \mathbf{JAJ'}, \tag{6.42}$$

where $\mathbf{J} = \mathbf{J_5 J_4}$ is again the product of the Jacobian matrices. The elements of these matrices may be defined as follows:

$$\mathbf{A} = \begin{pmatrix} \kappa_2 & \mathbf{Y'} \\ \mathbf{Y} & \mathbf{B}f_0^{-1} \end{pmatrix},$$

where $y_j = jT_j$,

$$\mathbf{B}(i, j) = T_i(1 + T_i), \quad i = j$$
$$= T_i T_j, \quad i \ne j,$$
$$\mathbf{J}_4^{-1}(i, j) = \{(i - 1)!\}^{-1}, \quad i = j$$
$$= T_{i-j}/j!, \quad i > j \geqslant 2$$
$$= 0, \text{ otherwise},$$

and

$$\mathbf{J}_5(i, j) = 1, \quad i = j = 1$$
$$= -\gamma_i/\tau_i, \quad 2 \leqslant i = j \leqslant s$$
$$= 1/\tau_{i-1}, \quad i + 1 = j = 3, \ldots, s + 1$$
$$= 0, \text{ otherwise}.$$

6.11.3 Use of the zero frequency

If the zero frequency is used in conjunction with the sample factorial cumulants, replace $\mathbf{W}$ in (6.40) by $\mathbf{W}^* = \begin{bmatrix} \mathbf{W} \\ \mathbf{0} \quad 1 \end{bmatrix}$ and use $\mathbf{\eta}^* = (\eta_0, \eta_1, \ldots, \eta_{s-1}, a)'$, a as defined in Table 6.3. The functions a derive principally from the expression $f_0 = f_0(\mathbf{\eta})$, although there is a certain arbitrariness about the final form chosen. The forms for a given in Table 6.3 are those used by Hinz and Gurland. Then, proceeding as before, we obtain

$$\Lambda = \mathbf{\Sigma}^* = \mathbf{J}_6 \mathbf{G} \mathbf{J}_6', \tag{6.43}$$

where

$$nG = \left(\begin{array}{c|c} n\Sigma & -f_0 JM \\ \hline -f_0 M'J' & f_0(1 - f_0) \end{array} \right),$$

$$J_6 = \left(\begin{array}{c|c} I & 0 \\ \hline D'\alpha & \partial\alpha/\partial f_0 \end{array} \right),$$

and D' is the operator $\left(\dfrac{\partial}{\partial\eta_0}, \ldots, \dfrac{\partial}{\partial\eta_{s-1}} \right)$.

The estimators follow from (6.40) as before.

Example 6.5 (Hinz and Gurland, 1967)

To find the estimators for the two-parameter Neyman Type A using the sample values $\{h_0, h_1, \ldots, h_{s-1}, a\}$. Define $\xi^* = (\kappa_{(1)}, \eta_1, \ldots, \eta_{s-1}, a)'$, where a is given in Table 6.3. We use $\theta' = (\lambda_1\lambda_2, \lambda_2)$ so that

$$W' = \left(\begin{array}{ccccc} 1 & 0 & 0 & \cdots\cdots & 0 \\ 0 & 1 & 1 & \cdots\cdots & 1 \end{array} \right).$$

In this case, $a = f_0\eta_1 \exp\{\eta_0/\eta_1(1 - P)\}$ where $P = \exp(-\eta_1)$, so that

$$J_6 = \left(\begin{array}{cc} I & 0 \\ D'\alpha & \alpha/f_0 \end{array} \right),$$

where $D'\alpha = (\alpha/\eta_1)(1 - P, 1 + \eta_0 P - \eta_0(1 - P)/\eta_1, 0)$.

Considering the special case where $s = 2$, we find that

$$J = \left(\begin{array}{ccc} 1 & 0 & 0 \\ -\kappa_{(2)}/\kappa_{(1)}^2 & 1/\kappa_{(1)} & 0 \\ 0 & -\kappa_{(3)}/\kappa_{(2)}^2 & 1/\kappa_{(2)} \end{array} \right) \left(\begin{array}{ccc} 1 & 0 & 0 \\ -1 & 1 & 0 \\ 2 & -3 & 1 \end{array} \right)$$

$$\times \left(\begin{array}{ccc} 1 & 0 & 0 \\ -2\mu_1' & 1 & 0 \\ -3\mu_2' + 6(\mu_1')^2 & -3\mu_1' & 1 \end{array} \right).$$

V, **M** and **G** are as defined above. By substituting the sample moments for the population parameters in these equations we may compute Σ^* and thus $\hat{\lambda}_2 = \hat{\theta}_2$, $\hat{\lambda}_1 = \hat{\theta}_1/\hat{\theta}_2$.

We note that the simple cases $\{\eta_0, \eta_1\}$ and $\{\eta_0, \alpha\}$ reduce to the use of the first two moments and the first moment and zero frequency, respectively (**W** or $\mathbf{W}^* = \mathbf{I}_2$ in these cases).

These MCS estimators are asymptotically fully efficient for $s \to \infty$, but it has been shown by Hinz and Gurland that even for $s = 2$ the resulting estimates are highly efficient. For example, when $\max(\lambda_1, \lambda_2) \leqslant 2$, the estimators in Example 6.5 are 99 per cent efficient or better. For the negative binomial the efficiency of the estimator $\{h_0, h_1, h_2, a\}$ is

$$\geqslant 98\%, \quad w = q/p \leqslant 2, \quad \text{any } k$$

$$\geqslant 93\%, \quad w \leqslant 5, \quad \text{any } k$$

$$\geqslant 88\%, \quad w \leqslant 10, \quad \text{any } k.$$

In general, for the distributions considered, the use of $\{h_0, h_1, a\}$ or the slightly preferable $\{h_0, h_1, h_2, a\}$ yields high efficiency in most parts of the parameter space.

Hinz and Gurland (1970) have further developed this approach to yield goodness-of-fit tests for the contagious distributions. Limited numerical studies suggest that these tests are more powerful than the classical χ^2 test.

6.12 Maximum likelihood (ML) estimators

ML estimators for generalised Poisson distributions have been considered by several authors, notably, for the negative binomial, Bliss and Fisher (1953); for the Neyman Type A, Neyman (1939), Shenton (1949) and Douglas (1955); for the Poisson–binomial, Sprott (1958), Shumway and Gurland (1960a, b) and Katti and Gurland (1962); and for the Poisson–Pascal, Shumway and Gurland (1960b) and Katti and Gurland (1961). A general study has been carried out by Martin and Katti (1965). For the estimation of two parameters in these models (that is, k is known for the Poisson–Pascal), these various results have been brought together by Sprott (1965) in

Theorem 6.3

If the ML equations can be written in the form

$$\left. \begin{array}{r} \Sigma \{n_r(r+1) f(r+1)/f(r)\} = n\bar{r} \\ c\hat{\theta}_1\hat{\theta}_2 = \bar{r} \end{array} \right\} \tag{6.44}$$

where n_r is the number of counts with r units observed, $n\bar{r} = \Sigma\, rn_r$, c is a function of the known parameters, $\hat{\theta}_1$ and $\hat{\theta}_2$ are the ML estimators of the unknown parameters θ_1 and θ_2, and prob $(R = r) = f(r|\theta_1, \theta_2)$, then the distribution must have a pgf of the form

$$G(z) = \phi(\delta) \sum_m \{1 - \gamma + \gamma z\}^m\, h(m)(-\delta)^m, \qquad (6.45)$$

where γ and δ are parameters. That is, the distribution must be expressible as a generalised binomial.

Proof. See Sprott (1965), pp. 338–42. For the converse, see
Exercise 6.11.

Example 6.6
　　The Neyman Type A has pgf

$$G(z) = \exp\left[\lambda_1\{\exp(\lambda_2 z - \lambda_2) - 1\}\right],$$

which may be written as

$$\exp(-\delta)\exp\{\delta^{(1-\gamma+\gamma z)}\},$$

where $\gamma = \lambda_2/\log \lambda_1$ and $\delta = \lambda_1$. Therefore,

$$f(r) = \sum_m p(m)\binom{m}{r}\gamma^r(1 - \gamma)^{m-r},$$

where
$$p(m) = e^{-\delta}\frac{(\log \delta)^m}{m}\sum_{j=0}^{\infty}\frac{j^m}{m!}.$$

Thus the Neyman Type A can be regarded as a generalised binomial and has ML equations of the form (6.44). A general result deriving from this example is that

Binomial $(m, \gamma)\underset{m}{\vee}$ general $A \sim$ Poisson $(m\gamma\delta)\underset{m}{\vee}$ general B,

where the density functions are

for general distribution A, 　$A_m\phi(\delta)\delta^m/m!$,

and for B, 　$a_m\phi(\delta)e^{-\delta m}$,

where 　$A_m = \sum_{i=0}^{\infty} a_i(-1)^i i^m$.

From these results, (6.44) can be seen to be the ML equations of the two-parameter generalised Poisson distributions listed. The general binomial $\vee$ general binomial, and the general binomial $\vee$ Poisson distributions of Khatri and Patel (see section 6.8), also have ML equations of form (6.44); the general binomial $\vee$ hypergeometric does not have ML equations of this form, however.

Having derived the likelihood equations, these may be solved iteratively and the information matrix obtained in the usual way (Sprott, 1965, pp. 344–5).

In particular cases, however, the ML equations may be further simplified, as the following example shows.

Example 6.7 (Shumway and Gurland, 1960a)

The Poisson–binomial has

$$f(r) = e^{-\lambda} \sum \binom{mt}{r} p^r q^{mt-r}/t!,$$

the sum taken over $t = 0, 1, \ldots$. If X is a Poisson variate with parameter $\omega = \lambda q^m$ then mX has rth non-central factorial moment

$$\mu_{(r)} = r!\, e^{-\omega} \sum \binom{mt}{r} \omega^r/t!$$

so that $\quad f(r) = e^{\omega-\lambda} (p/q)^r \mu_{(r)}/r!$

and $\quad (r + 1)\, f(r + 1)/f(r) = (p/q)M_r,$

where $\quad M_r = \mu_{(r+1)}/\mu_{(r)}.$

Thus the first ML equation reduces to

$$L(\hat{\omega}) = \Sigma\, n_r M_r - n\bar{r} = 0,$$

which can be solved for $\hat{\omega}$ by Newton's method, since the M_r are functions of ω only. To facilitate calculations, Shumway and Gurland give tables of M_r for $m = 2$. The same approach was earlier employed by Douglas (1955) for the Neyman Type A, a limiting case of the Poisson–binomial. Douglas also gives tables to assist calculations.

6.13   Estimators which use observed frequencies only

Consider a generalised Poisson with two unknown parameters. From (6.31), if $g(z) = \Sigma\, p_i z^i$ and $G(z) = \Sigma\, f_i z^i$, we see that

$$\left. \begin{array}{l} f_0 = \exp\{-\lambda(1 - p_0)\} \\[2mm] f_1 = \lambda p_1 \exp\{-\lambda(1 - p_0)\}, \text{ etc} \end{array} \right\} \tag{6.46}$$

If the probability that $r > 2$ is small, then we may reasonably restrict our attention to the values $r = 0$, 1 and $r \geqslant 2$. Reparametrising (6.46) so that $f_0 \equiv \theta_1$, $f_1 = \theta_2$, the ML estimators for the θ's are easily shown to be $\hat{\theta}_1 = n_0/n$, $\hat{\theta}_1 = n_1/n$, and we may then solve $\hat{f}_i = \hat{\theta}_{i+1}$, $i = 0$, 1 to obtain the ML estimators for the original parameters. Extension of this approach to k parameters, $n\hat{\theta}_j = n_j$, $j = 1, \ldots, k$ is immediate. This method was used to estimate the parameters of the Thomas distribution by Thomas (1949), and is particularly useful when the cost of counting items is high. The estimators based on γ_j in section 6.11 might be regarded as an extension of this idea.

6.14   Two examples

To show how the methods discussed in this chapter work in practice we consider two well-known data sets, drawn from Beall (1940) and McGuire *et al*. (1957).

In Table 6.5 the I and S values suggest that the Poisson–binomial should be used. This was fitted using (m_1, m_2), (m_1, f_0) and by maximum likelihood. The (m_1, f_0) is almost as good as the ML fit, since $r = 0$ accounts for 70 per cent of the observations. Where inefficient estimators have been used, the true distribution of the χ^2 goodness-of-fit statistic is a weighted sum of chi-square variables, rather than chi-square, but the figures quoted serve as a quide. The two-parameter hyper-Poisson is a reasonable fit, but not as good as the Poisson–binomial.

The I, S values in Table 6.6 are rather unsatisfactory because of the very skew nature of the distribution. However, the I, S values and the h_j and l_j plots suggest the use of either the negative binomial or the Poisson–Pascal, and the u_r ratios (see section 5.11) also suggest something like the negative binomial. From the table it is clear that the Neyman Type A is not a good fit, but that the other two provide very good fits, whatever the fitting method used. The better fit given by the l_j estimators reflects the greater stability of these sample statistics for this data set.

6.15   Independence of quadrats

When generalised Poisson distributions are fitted to data derived from areal sampling, it is assumed that there is independence between quadrats. In some investigations the quadrats, or points, are chosen sufficiently far apart for this assumption to be justified. In other situations, however, a grid may be laid over the study area so that it is exhaustively sampled. In such cases the assumption of independence must be checked.

Table 6.5

Observed and expected frequencies of Pyrausta nubilalis (European corn borer) in 1296 corn plants [Data from McGuire et al. (1957)]

No. of borers per plant	Observed frequency	Number of plants: fittings based on				
		Poisson–binomial ($m = 2$)*			Hyper-Poisson†	
		(m_1, m_2)	(m_1, f_o)	ML	ML	(m_1, m_2, f_o)
0	907	904·4	907	906·2	901·4	904·1
1	275	279·4	275·2	276·7	289·1	282·6
2	88	89·1	90·3	89·9	80·4	79·4
3	23	18·6	19·0	18·9	19·7	23·1
≥4	3	4·5	4·6	4·4	5·4	6·8
d.f.		2	2	2	2	2
χ^2		1·60	1·46	1·38	3·30	3·31
$P(\chi^2)$		0·45	0·48	0·50	0·19	0·19
$\tilde{\theta}$		—	—	—	2·091	2·779
$\tilde{\lambda}$		0·8296	0·7861	0·8752	6·519	8·890
ϕ		—	—	—	1·4378	1·4335
p		0·2474	0·2611	0·2321	—	—

$m'_1 = 0·410\,49$, $m_2 = 0·512\,05$, $m_3 = 0·6829$; $\tilde{f}_o = 0·699\,85$. $I = 1·247$, $S = 1·334$.

* From Shumway and Gurland (1960a)
† From Crow and Bardwell (1965)

Table 6.6 *Fitting a distribution of*

Data from

Number of beetles	Observed frequency	Fittings	
		Negative binomial using maximum likelihood	Neyman Type A using maximum likelihood
0	33	33·2	31·2
1	12	11·3	6·0
2	5	6·7	7·8
3	6	4·5 ⎫	7·2
4	6	3·2 ⎭	5·6
5	5	2·4 ⎫	4·0 ⎫
6	0	1·8 ⎬	2·8 ⎭
7	2	1·4 ⎭	1·9 ⎫
8	2	1·1 ⎫	1·3 ⎪
9	0	0·9 ⎪	0·8 ⎬
10	1	0·7 ⎪	0·5 ⎪
>10	2	2·8 ⎭	0·9 ⎭
χ^2		2·32	11·17
d.f.		3	4
$P(\chi^2)$		0·51	0·02
Estimates:		$k = 0·407$ $\hat{w} = 5·255$	$\hat{\lambda}_1 = 0·889$ $\hat{\lambda}_2 = 2·410$

$m'_1 = 2·133$, $m_2 = 13·636$, $m_3 = 137·52$, $\bar{f}_0 = 0·471\,43$, $I = 6·39$,
The curly brackets denote the groupings used in evaluating χ^2.

147

Colorado potato beetles, Leptinotarsa decemlineata
Beall (1940)

based on:

Negative binomial using (l_0, l_1, l_2, l_3)	Negative binomial using (h_0, h_1, a)	Poisson–Pascal using (h_0, h_1, h_2, a)
29·4	33·3	34·4
12·8	11·3	10·7
7·9	6·7	6·5
5·3	4·5 ⎱	4·4 ⎱
3·7 ⎱	3·2 ⎰	3·2 ⎰
2·7	2·4 ⎱	2·4 ⎱
2·0	1·8 ⎬	1·8 ⎬
1·5 ⎰	1·4 ⎰	1·4 ⎰
1·1 ⎱	1·1 ⎱	1·1 ⎱
0·8	0·9	0·9
0·6 ⎬	0·7 ⎬	0·7 ⎬
2·2 ⎰	2·7 ⎰	2·5 ⎰
1·73	2·35	2·52
3	3	2
0·63	0·50	0·28
$\hat{k} = 0\cdot546$	$\hat{k} = 0\cdot431$	$\hat{\lambda} = 4\cdot738$
$\hat{w} = 3\cdot914$	$\hat{w} = 4\cdot697$	$\hat{k} = 0\cdot093$
		$\hat{w} = 4\cdot715$

$S = 10\cdot1$; $l_1 = 0\cdot47$, $l_2 = 5\cdot01$, $l_3 = 2\cdot15$, $h_1 = 5\cdot39$, $h_2 = 8\cdot77$

Reproduced from Hinz and Gurland, 1967

This problem was first recognised by Student (1907), who calculated the correlation between each quadrat and its four neighbouring quadrats on a square lattice. However, the problem is really auto-correlative in nature, and a better statistic to use would be that of Cliff and Ord (1969). If x_i denotes that the value of variate X is quadrat i, put $z_i = x_i - \bar{x}$ and use the statistic

$$r = \left(\frac{n}{W}\right) \frac{\Sigma_{(2)} \, w_{ij} z_i z_j}{\sum\limits_{i=1}^{n} z_i^2} , \qquad (6.47)$$

where w_{ij} is the weight attached to the link between the ith and jth quadrats ($w_{ij} \geqslant 0$), $W = \Sigma_{(2)} \, w_{ij}$ and $\Sigma_{(2)} \equiv \sum\limits_{i=1}^{n} \sum\limits_{\substack{j=1 \\ i \neq j}}^{n}$. Under H_0:

independence between quadrats, the moments of r may be evaluated under either of two assumptions:

N: the x_i are independent drawings from a normal population,

or $\quad R$: the sampling frame is the set of $n!$ random permutations of the x_i.

With N, R as subscripts to denote the relevant assumption, we have

$$E_N(r) = E_R(r) = -(n-1)^{-1},$$

$$E_N(r^2) = (n^2 S_1 - nS_2 + 3W^2)/W^2(n^2 - 1),$$

and $\quad E_R(r^2) = [n\{(n^2 - 3n + 3)S_1 - nS_2 + 3W^2\} -$

$$- b_2\{(n^2 - n)S_1 - 2nS_2 + 6W^2\}]/W^2(n-1)^{(3)} .$$

In these equations, $\quad S_1 = \frac{1}{2} \Sigma_{(2)} \, (w_{ij} + w_{ji})^2,$

and $\quad S_2 = \sum\limits_{i=1}^{n} (w_{i.} + w_{.i})^2,$

where $w_{i.} = \sum\limits_{j=1}^{n} w_{ij}$ and $w_{.j} = \sum\limits_{i=1}^{n} w_{ij}$, while b_2 is the sample kurtosis coefficient m_4/m_2^2. Under appropriate conditions r may be shown to be asymptotically normally distributed, and approximations to the tail regions of the distribution for moderate n are suggested in Cliff and Ord (1970). That paper also gives an example of the use of r in testing for dependence between quadrats when the quadrats are grouped in different ways to try to reduce the dependence. Random processes in which there is dependence between quadrats but which are still generalised Poisson in form may be constructed (see Exercises 6.13 and 6.14).

When observations are taken over more than one time-period, however, even this restricted dependence is not allowable (see Ord, 1970).

EXERCISES

6.1   Use the results in Example 6.1 and equation (6.16) to derive MVU estimators for the distribution function of the negative binomial.

(Patil and Wani, 1966)

6.2   Find the pgf of the Poisson $\wedge$ beta distribution. Hence derive the the factorial cumulant generating function for the Gurland system of generalised distributions, and the factorial cumulants.

(Beall and Rescia, 1953; Gurland, 1958)

6.3   Find all the sets of conditions under which the Gurland distribution tends to the Neyman Type A (there are four).

(Gurland, 1958)

6.4   Find the pgf of the Poisson $\vee$ GPSD distribution and use this to derive the factorial cumulants for the negative binomial, Pólya–Aeppli, Neyman Type A, Poisson–binomial and Poisson–Pascal distributions.

6.5   Derive an expression for the pgf of the Poisson $\wedge$ Truncated Normal $[N(\mu, \sigma^2 \mid x > 0)]$ distribution. Show that for $\mu \gg \sigma^2$, this reduces to the Hermite distribution. Further show that the Normal $\vee$ Poisson is Hermite under appropriate conditions.

(Kemp and Kemp, 1965, 1966)

6.6   Show that the Neyman Type A may be represented as Poisson $\wedge$ Poisson, where the second Poisson has parameter λ_1 and is defined on a lattice of width λ_2.

(Gurland, 1957)

6.7   Show that Pascal $(k, q) \underset{k}{\wedge}$ gamma $\sim$ Pascal $\vee$ logarithmic.

(Gurland, 1957)

6.8   Derive the density function of the Poisson $\wedge$ lognormal (PL), and also find the factorial cumulants $\kappa_{(i)}$. To show the greater non-normality of this distribution, show that, when the first two moments of this distribution are equal to those of a negative binomial (NB) with index k, $\kappa_{(3)}\{PL\} \rightarrow 1 \cdot 5\, \kappa_{(3)}\{NB\}$ and $\kappa_{(4)}\{PL\} \rightarrow 2 \cdot 67\, \kappa_{(4)}\{NB\}$ for large k.

(Anscombe, 1950)

6.9   Use (6.31) to derive finite series expansions for the Neyman
Types B and C distributions.

6.10   If $$\mu'_{(s)} = \left.\frac{\partial\{h(z)\}}{\partial z^s}\right|_{z=1}$$

where $h(z)$ is defined by (6.35), establish the following relationships for the Khatri–Patel distributions. $\kappa_{(s)}$ and $m'_{(s)}$ denote the
sth factorial cumulant and moment respectively.

Type A: $\kappa_{(s)} = \mu'_{(s)}$

Type B: $\kappa_{(s)} = n\mu'_{(s)} - \sum\limits_{j=1}^{s-1}\binom{s-1}{j}\mu'_{(j)}\kappa_{(s-j)}$

Type C: $m'_{(s)} = \{c\mu'_{(s)} - \sum\limits_{j=1}^{s-1}\binom{s-1}{j}\mu'_{(j)}m'_{(s-j)}\}/h(1)$.

(Khatri and Patel, 1961)

6.11   Using (6.45), show that the maximum likelihood equations for
the class of generalised binomial distributions reduce to

$$n\overline{r} = \Sigma n_r(r+1)f(r+1)/f(r)$$

and $$\overline{r} = -\gamma\,\delta\phi'(\delta)/\phi(\delta).$$

Hence write down the ML equations for the Poisson–Pascal with
k known and formulate an iterative method of solution.

(Sprott, 1965)

6.12   Using the criteria suggested in section 6.9, select an appropriate
generalised Poisson distribution and fit it to the data given below.
Test the resulting goodness of fit.

No. of cells per square in a haemocytometer count

No. of cells	0	1	2	3	4	5
Observed no. of squares	213	128	37	18	3	1

$n = 400$, $m'_1 = 0.6825$, $m_2 = 0.8117$, $m_3 = 1.0876$.

(data from "Student", 1907)

6.13   Show that the pure birth process defined by the set of differential
equations

$$\frac{df_r}{dt} = \lambda f_{r-1} - \lambda f_r, \quad r = 0, 1, \ldots,$$

where $f_r \equiv f(r, t) = $ prob (r births by time t), $f(-1, t) = 0$ and
$f(0, 0) = 1$; is Poisson with parameter λt. Find a corresponding
set of equations for the generalised Poisson processes.

6.14　Consider a study area in which there are N quadrats and quadrat j is contiguous to the quadrats with indices in the set $J = \{i(1), \ldots i(k_j)\}$ where k_j is the number of quadrats contiguous to j. Show that the process defined by the set of differential equations

$$\frac{df(r, j)}{dt} = -\lambda \Delta_r f(r - 1, j) - 2 \rho k_j^{-1} \sum_{i \in j} \Delta_r f(r - 1, i)$$

$$r = 0, 1, \ldots, \quad j = 1, \ldots, N,$$

is a Poisson process with parameter $(\lambda + 2\rho)t$. Compare this result with that of Exercise 6.13. (Ord, 1970)

6.15　Find the pgf of the Poisson $(\lambda) \wedge$ uniform (on $[\alpha, \beta]$) distribution and hence determine the first two moments. Use the pgf to find the density function and hence determine the posterior distribution for λ. (Bhattacharya and Holla, 1965)

6.16　If r_i follows the generalised Poisson distribution with probability generating function $\exp [\lambda_i \{g(z) - 1\}]$, find the distribution of

$$S_n = \sum_{i=1}^{n} r_i.$$

Under what conditions does the distribution of S_n/n tend to normality as n tends to infinity? (Hint: Use Theorem 6.1 and Property M5 in section 4.2).

6.17　Find the first three moments of the "Short" distribution ($\sim$ Neyman Type A $\vee$ Poisson) and show that the I, S region for this distribution lies above that for the Neyman Type A. (Kemp, 1967a)

CHAPTER 7

MULTIVARIATE SYSTEMS OF DISCRETE DISTRIBUTIONS

7.1 Introduction

When sampling attributes, A or $\bar{A}$ (not A) say, two extensions are possible. The number of classes may be increased from two to $m + 1$, giving the set of multiple outcomes $\{A_0, A_1, \ldots, A_m\}$; alternatively, more than one attribute may be considered, giving the set of outcomes $\{AB, A\bar{B}, \bar{A}B, \bar{A}\bar{B}\}$ for example. The second situation is more truly "multivariate" and we reserve the label for this case. Formally, the multivariate model may be regarded as a multiple outcome model by considering each possible combination of attributes as an outcome, but there is clearly a difference of emphasis. Finally, a multivariate multiple outcome model may be formulated; realisations of such models can be represented as multiway contingency tables. These sampling models are formulated as urn schemes in section 7.2.

The generalised power-series distributions may also be extended to multivariate forms, and these are discussed in section 7.6. Multivariate contagious distributions are also of value as they enable the investigator to distinguish "true" and "apparent" contagion. This generally involves observing a given system over two or more non-overlapping time-periods, and is therefore rather different in conception from the sampling models.

7.2 The classical distributions

The Pólya—Eggenberger urn scheme described in section 5.4 may be extended to contain M_i balls of colour i, $i = 0, 1, \ldots, m$, N balls in all. If a ball of colour i is drawn, that ball and c further balls of the same colour are replaced in the urn and the next drawing made. The process continues until terminated by an appropriate stopping rule.

For direct sampling, when n drawings are made, the probability of the sample configuration $\mathbf{r} = \{r_0, r_1, \ldots, r_m\}$ is

$$\text{prob}(\mathbf{R} = \mathbf{r}) = f(\mathbf{r}) = \binom{n}{r_1, \ldots, r_m} \{\Pi(M_i)^{(r_i, c)}\}/(N)^{(n, c)}, \quad (7.1)$$

where the product is taken over $i = 0, 1, \ldots, m$,

$$(A)^{(x,c)} = A(A + c) \ldots \{A + (x - 1)c\},$$

and $\begin{pmatrix} n \\ r_1, \ldots, r_m \end{pmatrix} = n!/\{r_0! r_1! \ldots r_m!\}.$

A variety of inverse sampling schemes may be formulated, sampling ceasing when $\sum_I r = k$, I being a subset of $\{0, 1, \ldots, m - 1\}$. The usual model assumes that sampling ceases when $r_0 = k$, yielding

$$\text{prob } (\mathbf{R} = \mathbf{r}) = f(\mathbf{r}) = \begin{pmatrix} k + S - 1 \\ r_1, \ldots, r_m \end{pmatrix} \{\Pi(M_i)^{(r,c)}\}/N^{(k+S,c)}, \quad (7.2)$$

where $S \equiv S(\mathbf{r}) = \sum_{i=1}^{m} r_i.$

Both (7.1) and (7.2) may be represented by the set of difference equations

$$\Delta_i f(\mathbf{r}) = \frac{a_{i0} + a_{i1} r_i + a_{i2} S}{b_{i0} + b_{i1} r_i + b_{i2} r_i S} f(\mathbf{r}), \quad (7.3)$$

where Δ_i denotes differencing with respect to the ith variate and the a_{ij}, b_{ij} are functions of the parameters. The set of equations (7.3) is a special case of the more general set

$$\Delta_i f(\mathbf{r}) = \frac{\text{linear form in } r_0, r_1, \ldots, r_m}{\text{quadratic in } r_0, r_1, \ldots, r_m}. \quad (7.4)$$

This extended form also yields density functions which are expressible as hypergeometric functions, but only the more restricted (7.3) is considered here.

A Bayesian derivation of a sub-class of (7.3) is now given, due to Särndal (1965).

Consider K balls, indistinguishable except for being coloured in one of $(m + 1)$ ways. Then, the Bose—Einstein prior distribution for different configurations is (Feller, 1957, p. 18)

$$P(K_1, K_2, \ldots, K_m \mid \sum_{j=1}^{m} K_j \leq K) = \begin{pmatrix} K + m \\ m \end{pmatrix}^{-1}.$$

If a first sample, of size N, is drawn from this population, with frequencies $\mathbf{N} = (N_1, \ldots, N_m)$, $\sum N_j \leq N$, the probability of $\mathbf{N}$, given $\mathbf{K} = (K_0, K_1, \ldots, K_m)$, is

$$P(\mathbf{N} \mid \mathbf{K}) = \left\{ \prod_{i=0}^{m} \begin{pmatrix} K_i \\ N_i \end{pmatrix} \right\} / \begin{pmatrix} K \\ N \end{pmatrix}. \quad (7.5)$$

Now draw a second sample of size n with frequencies

$$\mathbf{r} = \{r_0, r_1, \ldots, r_m\};$$

then we have

$$P(\mathbf{r} \mid \mathbf{N}, \mathbf{K}) = \left\{ \prod_{i=0}^{m} \binom{K_i - N_i}{r_i} \right\} \bigg/ \binom{K - N}{n}. \tag{7.6}$$

It follows that the conditional distribution of $\mathbf{r}$, given $\mathbf{N}$, after summing over all possible $\mathbf{K}$, is

$$P(\mathbf{r} \mid \mathbf{N}) = \left\{ \prod_{i=0}^{m} \binom{N_i + r_i}{r_i} \right\} \bigg/ \binom{N + m + n}{n}. \tag{7.7}$$

It is easily seen that (7.1) reduces to (7.7) for $c = 1$. An alternative specification for (7.7) is the mixture

multinomial $\wedge$ multivariate beta (Dirichlet);

the multinomial being given by (7.1) for $c = 0$, that is

$$f(\mathbf{r}) = \binom{n}{r_1, \dots, r_m} q^{r_0} \prod_{i=1}^{m} p^{r_i}, \tag{7.8}$$

while the Dirichlet distribution is of the form (cf. equation (3.2) in section 3.2)

$$f(\mathbf{y}) = \Gamma(\Sigma \, a_i) \prod_{i=0}^{m} \{ y_i^{a_i} / \Gamma(a_i) \}, \tag{7.9}$$

with $\sum_{i=0}^{m} y_i = 1$. We therefore refer to (7.7) as the Dirichlet–multinomial, the multiple analogue of the beta–binomial.

A similar Bayesian interpretation may be given for (7.2) when $c = 1$, yielding the

negative multinomial $\wedge$ Dirichlet or Dirichlet–Pascal distribution,

from the negative multinomial

$$f(\mathbf{r}) = \binom{k + S - 1}{r_1, \dots, r_m} q^k \prod_{i=1}^{m} p_i^{r_i}. \tag{7.10}$$

The distributions defined by (7.1) and (7.2) may be summarised as follows:

	direct sampling	*inverse sampling*
$c = -1$	multiple hypergeometric	multiple negative hypergeometric
0	multinomial	negative multinomial
$+1$	Dirichlet–multinomial	Dirichlet–Pascal.

We note that the equivalence of the beta-binomial and the negative hypergeometric is lost when $m > 1$. A wide variety of limiting forms exists for

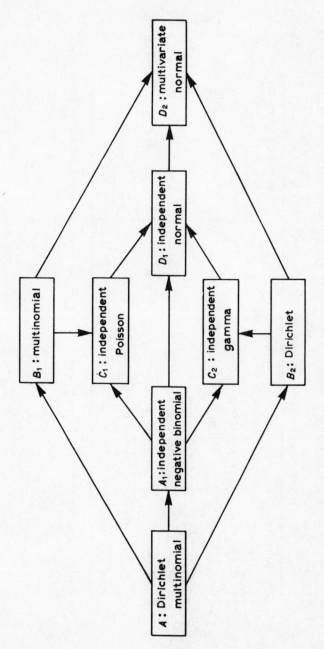

Fig.7.1 The relationships stemming from the Dirichlet–multinomial
(Adapted from Särndal, 1965)

Table 7.1

Conditions for some limiting relations displayed in Fig. 7.1

Limiting relation	Conditions ($i = 0, 1, \ldots, m$)	Remaining parameters ($i = 0, 1, \ldots, m$)
$A \to B_1$	$N \to \infty$, $p_i = M_i/N$	n, p_i
$A \to A_1$	$r_0, M_0, n, N \to \infty$, $p = N/(N+n)$	p, M_i $(i \neq 0)$
$n^m A \to B_2$	$n \to \infty$, $r_i/n \to y_i$	M_i
$B_1 \to C_1$	$N, M_0 \to \infty$, $p_i \to 0$, $M_i > 0$	M_i
$A_1 \to C_1$	$M_i \to \infty$, $M_i(1-p)/p \to \theta_i$, $i \neq 0$	θ_i $(i \neq 0)$
$p^{-m} A_1 \to C_2$	$r_i \to \infty$, $p \to 0$, $r_i p/(1-p) \to y_i$	M_i
$N^{-m} B_2 \to C_2$	$N, M_0 \to \infty$, $y_i \to 0$, $Ny_i \to x_i$, $i \neq 0$	M_i

The conditions under which C_1, $C_2 \to D_1$ and B_1, $B_2 \to D_2$ are well known and are not repeated here.

these distributions; Fig. 7.1 shows the relationships developed by Särndal (1965) for the Dirichlet–multinomial; a summary is given in Table 7.1. For other distributions see Exercise 7.1.

7.3 Multivariate classical forms

So far, multiple outcome, rather than truly multivariate, models have been considered. However, the link is easily forged. Apart from a constant, the $f(\mathbf{r})$ are the terms of the multivariate hypergeometric function

$$F(a, b_1, \ldots, b_m; d; 1, \ldots, 1), \qquad (7.11)$$

where a, d and the b_i are suitably chosen functions of the parameters. The series expansion of (7.11) converges if $d(d - a - \Sigma b_i - 1) > 0$. The probability generating function is therefore proportional to

$$F(a, b_1, \ldots, b_m; d; z_1, \ldots, z_m), \qquad (7.12)$$

while the factorial cumulant generating function is the logarithm of (7.12), when the $\{z_i\}$ are replaced by $\{1 + u_i\}$.

If $m = 2^p - 1$, suitable reparametrisation of $z_{p+1}, \ldots, z_m$ yields the pgf of the p-fold multivariate form directly. For example, if $m = 3$, $p = 2$ and $z_3 = z_1 z_2$. The general form of the bivariate density function is therefore

$$f(r_1, r_2) = \sum_{s=0}^{y} f(r_1 - s, r_2 - s, s), \qquad (7.13)$$

where $y = \min(r_1, r_2)$.

Example 7.1

The multinomial distribution, with $m = 3$, has pgf

$$(q + p_1 z_1 + p_2 z_2 + p_3 z_3)^n.$$

Putting $z_3 = z_1 z_2$, $p_3 \equiv p_{12}$, we have the pgf of the bivariate binomial

$$(q + p_1 z_1 + p_2 z_2 + p_{12} z_1 z_2)^n.$$

From (7.13) it follows that the density function for the bivariate binomial is

$$f(r_1, r_2) = \sum_{s=0}^{y} \binom{n}{r_1 - s, r_2 - s} q^{r_0 + s} p_1^{r_1 - s} p_2^{r_2 - s} p_{12}^s.$$

The complete list of distributions, with their factorial cumulant generating functions, first two factorial cumulants, and references to the literature, are given in Table 7.2. When observations are classified

Table 7.2 *Multivariate*

Distribution	Factorial cgf $(u_i = z_i - 1)$
Poisson (MU: Krishnamoorthy, 1951)	$\sum \lambda_i u_i$
Multinomial (MU: Wishart, 1949; Krishnamoorthy, 1951)	$n \ln (q + p_1 u_1 + \ldots + p_m u_m)$
Negative multinomial (MU: Wishart, 1949; Wiid, 1957/8; Sibuya *et al.*, 1964)	$-k \ln (1 - q_1 u_1 - \ldots - q_m u_m)$
Logarithmic (MO, MU: Patil and Bildikar, 1967)	$\ln \{-\ln (1 - q_1 u_1 - \ldots - q_m u_m)\}$
Hypergeometric $[c = -1]$ (MO: Pearson, 1924; Steyn, 1951; MU: Ram, 1954, 1955, 1956; Steyn, 1955, 1957)	$\ln \{F(-n, M_1/c, \ldots, M_m/c; -M_0 c^{-1} - n; u_1, \ldots, u_m)\}$
Dirichlet–multinomial $[c = 1]$ (MO: Ishii and Hayakawa, 1960; MU: Steyn and Wiid, 1958)	$p_i = M_i/N, M_0 = N - \sum_{i=1}^{m} M_i$
Negative hypergeometric $[c = -1]$ (MO: Steyn, 1951; MU: Steyn, 1957)	$\ln \{F(k, M_1/c, \ldots, M_m/c; Nc^{-1} + k + 1; u_1, \ldots, u_m)\}$
Dirichlet–Pascal $[c = -1]$ (MO: Mosimann, 1962, 1963; Särndal, 1964, 1965; MU: Steyn and Wiid, 1958)	$q_i = M_i/N; Np = M_0 - c$

Definitions: $p = 1 - \sum_{i=1}^{m} q_i$; $q = 1 - \sum_{i=1}^{m} p_i$; $a = \{-\ln p\}^{-1}$; $t_i = p_i + p_{12}$; $v_i = q_i + q_{12}$;

Notes: MO denotes multiple outcome; MU denotes multivariate.

in a p-way table (containing 2^p cells for binary data), the sample numbers in each cell provide natural and efficient estimates for the population parameters. However, the general situation, when only marginal totals are available, is more complex. We return to this problem in section 7.5.

discrete distributions

| Factorial cumulants | | | | |
| Multiple outcome | | Bivariate | | |
$\kappa((1, 0, \ldots, 0)) = \mu_i$	$\kappa((\delta, 2-\delta, 0, \ldots 0))$ $[\delta = 0, 1]$	$\kappa((1, 0))$	$\kappa((2, 0))$	$\kappa((1, 1))$
λ_i	0	$\lambda_i + \lambda_{12}$	0	λ_{12}
np_i	$-np_i p_j$	nt_i	$-nt_i^2$	$n(p_{12} - t_1 t_2)$
kq_i/p	$kq_i q_j/p^2$	kv_i/p	kv_i^2/p^2	$k(pq_{12} + v_1 v_2)/p^2$
$\alpha q_i/p$	$\alpha(1-\alpha)q_i q_j/p^2$	$\alpha v_i/p$	$\dfrac{\alpha(1-\alpha)v_i^2}{p^2}$	$\alpha\{(1-\alpha)v_1 v_2 + pq_{12}\}/p^2$
np_i	$-Enp_i p_j,\ i \neq j$ $np_i(E - Ep_i - 1),$ $\quad i = j$	nt_i	$nt_i(E - Et_i - 1)$	$np_{12} - nEt_1 t_2$
kq_i/p	$kDq_i q_j/p^2,\ i \neq j$ $kq_i(Dq_i + Dp - p)/p^2,$ $\quad i = j$	kw_i/p	$kw_i(Dw_i + Dp - p)/p^2$	$k(pq_{12} + Dw_1 w_2)/p^2$

$w_i = (M_i + M_{12})/N$; $D = (M_0 + kc - c)/(M_0 - 2c)$; $E = (N + nc)/(N + c)$.

7.4 Properties of the classical distributions

Recurrence relations for the densities may be obtained directly for the multiple outcome distributions from the difference equations (7.3) and (7.4). For the bivariate densities, useful general relationships may be obtained by evaluating

$$r_1 \, f(r_1, \, r_2) \; \equiv \; af(r_1 - 1, \, r_2) + bf(r_1 - 1, \, r_2 - 1) \,,$$

making suitable parameter adjustments. For the bivariate hypergeometric, for example, we find that

$$Nr_i \, f(r_1, \, r_2 \, | M_1, \, M_2, \, M_{12}, \, N, \, n)$$
$$= \; nM_1 f(r_1 - 1, \, r_2 \, | M_1 - 1, \, M_2, \, M_{12}, \, N - 1, \, n - 1) \, +$$
$$+ \; nM_{12} f(r_1 - 1, \, r_2 - 1 \, | M_1, \, M_2, \, M_{12} - 1, \, N - 1, \, n - 1). \qquad (7.14)$$

Recurrence relations for limiting forms of the hypergeometric can be found directly from (7.14); that equation also suggests the form of the recurrence relation for the m-variate case. We now list some relationships for moments and cumulants, noting that the ordering of the variables is irrelevant, so that the expressions are valid for all i, $1 \leqslant i \leqslant m$.

Property MD1 (Steyn, 1951). Distributions with factorial moment generating functions proportional to

$$F(a, \, b_1, \ldots, \, b_m; \, d; \, u_1, \ldots, \, u_m)$$

have non-central factorial moments

$$\mu'((j_1, \ldots, j_m)) \; = \; \frac{(-1)^J (a + J - 1)^{(J)} \Pi (b_i + j_i - 1)^{(j_i)}}{(a - d + J + \Sigma \, b_i - 1)^{(J)}}, \qquad (7.15)$$

where $J = \Sigma \, j_i$, and all sums and products are taken over $i = 1, \ldots, m$. The double bracket is used to denote *factorial* moments. From (7.15), we have the recurrence relation

$$\mu'((j_1, \ldots, j_i + 1, \ldots, j_m)) \; = \; \frac{-a(b_i + j_i)}{a + d + J + \Sigma \, b_i} \; \mu'((j_1, \ldots, j_m)). \qquad (7.16)$$

Property MD2 (Wishart, 1949). The set of distributions consisting of the multinomial, the negative multinomial and their limiting forms has cumulant generating function of the typical form

$$\psi \; \equiv \; \psi(1 + \theta_1 z_1 + \ldots + \theta_m z_m), \; z_i \; = \; \exp{(t_i)};$$

that is, in the terminology of Joshi and Patil (1970), the class is sum-symmetric (see section 7.6). It follows that the factorial cumulants, $\kappa((j_1, \ldots, j_m))$ obey the general relationship

$$\kappa((j_1, \ldots, j_i + 1, \ldots, j_m)) \; = \; \theta_i \, D_i \, \kappa((j_1, \, \ldots, \, j_m)), \qquad (7.17)$$

where $D_i = \partial / \partial \theta_i$.

Property MD3 (Wishart, 1949). To evaluate the cumulants of the p-variable multivariate analogue of the class considered in MD2, we recall that the cgf is unchanged except that $z_{p+1} = z_1 z_2$, $z_{p+2} = z_1 z_3, \ldots, z_m = z_1 z_2 \ldots z_p$. Therefore, instead of a single term in z_i in (7.17), we must pick up every term containing z_i. Putting

$$D_{ij\ldots} = \frac{\partial}{\partial \theta_{ij\ldots}},$$

we find that

for $p = 2$, $\quad \kappa(j_1 + 1, j_2) = (\theta_1 D_1 + \theta_{12} D_{12}) \kappa(j_1, j_2)$, $\hspace{1cm}$ (7.18)

for $p = 3$, $\quad \kappa(j_1 + 1, j_2, j_3) = (\theta_1 D_1 + \theta_{12} D_{12} + \theta_{13} D_{13} +$

$$+ \theta_{123} D_{123}) \kappa(j_1, j_2, j_3),\hspace{1cm} (7.19)$$

and so on, for higher p. If some $\theta_{ij\ldots}$ are zero, the corresponding $D_{ij\ldots}$ are omitted from the expressions.

Example 7.2

The bivariate binomial discussed in Example 7.1 has

$$\theta_* = p_*/q, * \equiv 1, 2, 12,$$

whence

$$\kappa(1, 0) = n(p_1 + p_{12}) = nt_1 = n(\theta_1 + \theta_{12})/(1 + \theta_1 + \theta_2 + \theta_{12}).$$

Applying (7.18), we see that

$$\kappa(2, 0) = n(\theta_1 D_1 + \theta_{12} D_{12}) \{(\theta_1 + \theta_{12})/(1 + \theta_1 + \theta_2 + \theta_{12})\}$$
$$= n(1 + \theta_2)(\theta_1 + \theta_{12})/(1 + \theta_1 + \theta_2 + \theta_{12})^2$$
$$= nt_1(1 - t_1).$$

Similarly, $\kappa(1, 1) = (\theta_2 D_2 + \theta_{12} D_1) \{\kappa(1, 0)\}$, which reduces to $n(p_{12} - t_1 t_2)$. These values may be checked from Table 7.2.

Useful recurrence relations for the moments and cumulants of the whole multivariate class are not available, but moments of lower orders could be derived from the general expansion

$$E\{(r_a + r_b + \ldots + r_x)^J\} = \Sigma \binom{J}{j_b, \ldots, j_x} \mu'(j_a, j_b, \ldots, j_x),$$

$$(7.20)$$

the sum being over all sets $\{j_a, j_b, \ldots, j_x \mid \Sigma j_i = J\}$. The subscripts $a, b, \ldots, x$ could be chosen to correspond to a particular variate; that is, all subscript groups containing a particular subscript: the subset $\{1; 12, 13, \ldots, 1p; 123, \ldots\}$, for example.

Property MD4 (Steyn, 1955). If a multiple outcome distribution has a pgf proportional to (7.12) then any subset of $k \leqslant m$ of the random variables has a pgf of the same form. This follows from the result

$$F(a, b_1, \ldots, b_m; d; z_1, \ldots, z_k, 1, \ldots, 1)$$
$$\propto F(a, b_1, \ldots, b_k; d - B; z_1, \ldots, z_k), \qquad (7.21)$$

where $B = \sum_{i=k+1}^{m} b_i$.

Property MD5 (Ram, 1955; Steyn, 1955). Let $\phi \equiv \phi(z_1, \ldots, z_m)$ denote the probability generating function of the class specified by equation (7.3). Then ϕ is given by (7.12), apart from a constant term, and satisfies the equation

$$(1 - z_j) \sum_{i=1}^{m} z_i D_i D_j \phi + \{d - (1 + a + b_j)z_j\}D_j \phi - b_j \sum_{i \neq j} z_i D_i \phi = ab_j\phi,$$
$$j = 1, \ldots, m, \qquad (7.22)$$

where $D_j = \partial/\partial z_j$. By the methods used in section 3.2 we may prove the discrete analogue of Theorem 3.1.

Theorem 7.1

The regression function for r_1 on $r_2, \ldots, r_m$ is given by

$$\tilde{r}_1 = \frac{Q(r_2, \ldots, r_m)}{R(r_2, \ldots, r_m)} \equiv \frac{Q}{R}, \qquad (7.23)$$

where Q and R are given by

$$[D_1\{R(D_2, \ldots, D_m)\phi\}] = [Q(D_2, \ldots, D_m)\phi], \qquad (7.24)$$

evaluated at $z_i = 1$, for all i. From (7.22) we find that, for all $z_i = 1$,

$$[(d - 1 - a - b_1)(D_1\phi) - ab_1\phi - b_1 \sum_{i \neq 1} D_i \phi] = 0,$$

so that the regression function is linear, being

$$(d - a - b_1 - 1)\tilde{r}_1 = b_1 a + b_1 \sum_{i \neq 1} r_i.$$

Property MD6 (Steyn, 1957). From Theorem 3.2 it may be shown that the bivariate distributions have linear regression functions of the form

$$(b_2 + b_{12})(d - a - b_1 - 1)\tilde{r}_1 = (b_2 + b_{12})ab_1 + \{b_1 b_2 + b_{12}(d - a - 1)\}r_2. \qquad (7.25)$$

For $p \geqslant 3$, however, the regression functions are linear only under special and increasingly restrictive conditions (Steyn and Wiid, 1958). Mahamunulu (1967) gives a detailed discussion of the conditions for the multivariate Poisson.

Property MD7. Although the univariate distributions are unimodal (or have adjacent twin modes), the multivariate distributions may be multimodal. A practical search procedure for the multinomial and multiple hypergeometric is given by Finucan (1964).

Property MD8 (extension of Young, 1967). To test the null hypothesis that the parameter for each of the m classes is the same ($H_0 : M_i = M$, $i = 1, \dots , m$) the mean deviation is sometimes used. That is $e_m = \sum_{j=1}^{m} |r_j - nm^{-1}|$, where n denotes sample size and r_j the number observed in the jth class. The expectation of e_m is the sum, over $1 \leqslant j \leqslant m$, of the expectations given in Table 5.2, while the evaluation of var (e_m) requires consideration of the bivariate densities. Young gives var (e_m) for the multinomial.

7.5 Estimation of parameters

It is not difficult to devise reasonable estimators for the multiple outcome models, but the multivariate models pose more severe problems.

For the multinomial and negative multinomial, of order $(m + 1)$ with known indices, there are m independent parameters and the ML estimators are the m sample proportions. When the index k is unknown in the negative multinomial, its ML estimator is the solution of an equation involving a digamma function (see Exercise 7.4).

In general, moment estimators, using the first- and second-order factorial moments, are easily derived from Table 7.2 (pages 158–9). However, Holgate (1964) showed for the bivariate Poisson (when $\lambda_1 = \lambda_2 = \lambda$) that the efficiency of the moment estimators drops sharply as λ_{12}/λ increases. The situation is unlikely to improve when $\lambda_1 \neq \lambda_2$, and similar results may be expected for other bivariate distributions. ML estimators for the bivariate Poisson, derived by Holgate, are given in Exercise 7.5.

For the hypergeometric series distributions, with $c \neq 0$ in Table 7.2, ML estimators are virtually intractable. However, Mosimann (1962, 1963) has developed a useful moments method for the multiple outcome distributions, which we now discuss.

There are $(m + 1)$ unknown parameters, which we refer to as $a' = (a_1, a_2, \dots , a_m)$ and g. Thus we have:

	a_i	g	θ
Hypergeometric } Dirichlet—multinomial }	M_i/N	$(N + nc)/(N + c)$	n
Negative hypergeometric } Dirichlet—Pascal }	$M_i/(M_0 - c)$	$(M_0 + kc - c)/(M_0 - 2c)$	k

where θ denotes the (known) index. It follows directly that the estimators

$$\tilde{a}_i = \bar{r}_i/\theta, \quad i = 0, 1, \ldots, m, \tag{7.26}$$

are unbiased. Further, the covariances are of the form

$$\text{cov}(r_i, r_j) = -\theta g a_i a_j, \quad i \neq j,$$

from which an estimate of g could be obtained. We suggest

$$\tilde{g} = \theta(1 - \tilde{a}_0)/\{\tilde{a}_0 \tilde{a}'\tilde{\Sigma}^{-1}\tilde{a}\}, \tag{7.27}$$

where $\tilde{\Sigma}$ is the sample $(m \times m)$ variance—covariance matrix for the a_i. Equations (7.26) and (7.27) may then be used to give estimates of the original parameters.

For truly multivariate situations the picture is less rosy as a p-variate distribution will have 2^p parameters ($2^p - 1$ if $g = 1$). Thus, third-order moments are required for $p \geqslant 4$, fourth-order for $p \geqslant 7$ and so on. The methods available for handling large numbers of parameters are still primitive, although the recent papers by Sprott (1970) and Kalbfleisch and Sprott (1970) represent a step forward.

The Poisson and both binomial forms are all multivariate power series distributions, and we now study this class.

7.6 Multivariate generalised power series distributions (MGPSD)

The multivariate analogues of the power series distributions considered in Chapter 6 were first outlined by Khatri (1959), but the major part of the work in this area first appeared in Patil (1965a).

The MGPSD has a density function of the form

$$f(\mathbf{r}) \equiv f(r_1, \ldots, r_m \mid \theta_1, \ldots, \theta_m) = a(\mathbf{r})\theta_1^{r_1} \ldots \theta_m^{r_m}/A(\boldsymbol{\theta}), \tag{7.28}$$

the range, $\mathbf{T}$, being the m-fold Cartesian product of the range of each variate, R_i. Several properties are now listed, taken from Patil's paper.

Property MG1. On inspection, it is seen from (7.28) that the recurrence relation

$$f(\mathbf{r} + \mathbf{j}) = \frac{a(\mathbf{r} + \mathbf{j})}{a(\mathbf{r})} \theta_1^{j_1} \ldots \theta_m^{j_m} f(\mathbf{r}) \tag{7.29}$$

holds for all $j \geqslant 0$. Further, if a multivariate discrete distribution obeys (7.29), then it must be a MGPSD.

Property MG2. Any subset of k ($\geqslant 1$) of the m variates with density function (7.28) has a k variate MGPSD. Further, the conditional distribution of the remaining $(m - k)$ variates, given the values of the selected k variates, is also a MGPSD.

Property MG3. The probability generating function is

$$A(\phi)/A(\theta), \tag{7.30}$$

where ϕ has elements $\phi_j = \theta_j z_j$.

Property MG4. Let $\kappa(s)$, $s' = (s_1, \ldots, s_m)$ denote the cumulant of order s. If $s + I_j = (s_1, \ldots, s_j + 1, \ldots, s_m)$,

then

$$\kappa(s + I_j) = \theta_j \frac{\partial \kappa(s)}{\partial \theta_j}. \tag{7.31}$$

Patil demonstrated that a distribution is MGPSD if and only if (7.31) holds. Recurrence relations may also be constructed for central and non-central moments, see Exercise 7.6.

Property MG5. The sample means, $\bar{r}_j$, $j = 1, \ldots, m$ provide the ML estimators for the θ_j. The m equations to be solved are

$$\bar{r}_j = \kappa(I_j), \quad j = 1, \ldots, m. \tag{7.32}$$

Property MG6. A sample of size N from a GPSD takes on values in the range $N[T]$, T being the range of values for a single observation. A necessary and sufficient condition for the function

$$g(\theta) = \theta_1^{v_1} \ldots \theta_m^{v_m}$$

to be minimum variance unbiased (MVU) estimable is that the vector $v' = (v_1, \ldots, v_m)$ should be contained in $N[T]$. This eliminates negative and fractional powers. From the definition, it follows that if $g(\theta)$ is MVU estimable for $N = N_0$, it remains so for all $N > N_0$.

Example 7.3

The multinomial is a MGPSD and its density may be written as

$$f(r) = \frac{n!}{(n - \Sigma r_i)!} \prod_{i=1}^{m} \left(\frac{\theta_i^{r_i}}{r_i!} \right).$$

For a single "observation" (n items constitute one "observation"),

166

only $g(\theta)$ for which $v^* = \sum_{i=1}^{m} v_i < n$ are MVU estimable functions.

Given N "observations", the number of items which may possess an attribute lies in the range $[0, Nn]$, and $g(\theta)$ is MVU estimable if $v^* < Nn$. The MVU estimator is, in fact,

$$\left.\begin{array}{c} \{\prod_{i=1}^{m} (r_i)^{(v_i)} \} / (Nn)^{(v^*)}, \quad \text{if } \mathbf{r} \geqslant \mathbf{v} \\[2mm] \text{zero, otherwise,} \end{array}\right\} \tag{7.33}$$

where $(a)^{(b)} = a(a - 1) \ldots (a - b + 1)$. The efficient estimation of general functions of the parameters has been discussed, for the multinomial, by Rao (1957, 1958) and Bhat and Kulkarni (1966).

Property MG7 (Joshi and Patil, 1970). If $A(\theta)$ in (7.28) may be written as $B(\theta^*)$, that is, as a function of $\theta^* = \Sigma \theta_i$ only, the MGPSD is said to be *sum-symmetric*. For such distributions, functions of $\theta_j^{v_j}$, $j = 1$, ..., m are MVU estimable if and only if $\Sigma \theta_j^{v_j}$ is MVU estimable. It follows that the probability generating function is of the form

$$B(\phi^*)/B(\theta^*), \quad \text{where } \phi^* = \Sigma \theta_j z_j.$$

Many of the more important sum-symmetric distributions also obey (7.3), and properties MD hold in these instances. For further properties, see Joshi and Patil's paper.

7.7 Models derived from the bivariate Poisson

From Table 7.2, we know that the bivariate Poisson law has the probability generating function (on redefining the parameters),

$$P(z_1, z_2) = \exp \{(\lambda_1 - \mu)(z_1 - 1) + (\lambda_2 - \mu)(z_2 - 1) + \mu(z_1 z_2 - 1)\}. \tag{7.34}$$

That is, given three independent Poisson variates U_1, U_2 and U_3 with means $\lambda_1 - \mu$, $\lambda_2 - \mu$ and μ respectively, we define the correlated variates

$$R_1 = U_1 + U_3, \quad R_2 = U_2 + U_3. \tag{7.35}$$

This construction provides the basis for the extensions which we now consider.

Bivariate generalised Poisson distributions, or "real contagion" models, are formed, as in section 6.6, on replacing z_i by $g_i(z_i)$, where g_i denotes the pgf of the generalising distribution. The term $z_1 z_2$ could be replaced by $g_1(z_1) g_2(z_2)$ or by $g_3(z_1 z_2)$. The use of g_3 preserves the interpretation (7.35) given above. Thus, two distinct bivariate generalised models are available:

I (correlated) bivariate Poisson $\vee$ two independent generalisers (with $g_1 g_2$).

II (correlated) bivariate Poisson $\vee$ three independent generalisers (with g_3).

These models are identical only when $\mu = 0$. To these a third model may be added, when a single variate gives rise to two (correlated) counts, that is

III (univariate) Poisson $\vee$ (correlated) bivariate generaliser.

Example 7.4

For the Poisson $\vee$ logarithmic, with $g_i = \alpha_i \ln (1 - q_i z_i)$, $\alpha_i = \{- \ln(1 - q_i)\}^{-1}$, the three models have pgf's

I $h_1(z_1) h_2(z_2) \exp (\mu g_1 g_2)$, where h_i denotes the pgf of a univariate negative binomial with parameters q_i, k_i and $k_i = \alpha_i(\lambda_i - \mu)$, $i = 1, 2$.

II $h_1(z_1) h_2(z_2) h_3(z_1 z_2)$, where h_3 also represents a NBD, with parameters q_3, $k_3 = \alpha_3 \mu$.

III the bivariate LSD has pgf as given in Table 7.2, so that the generalised distribution has pgf

$$(1 - q_1 z_1 - q_2 z_2 - q_{12} z_1 z_2)^{-\lambda \alpha}(1 - q_1 - q_2 - q_{12})^{\lambda \alpha},$$

where λ is the parameter of the univariate Poisson and $\alpha = \{- \ln (1 - q_1 - q_2 - q_{12})\}^{-1}$.

The corresponding set of alternatives for the Poisson $\vee$ Poisson is given by Holgate (1966). Model III is the most commonly used, and for the Poisson $\vee$ LSD corresponds to the compound model with a single gamma compounder. That is, given a bivariate Poisson with parameters $(\lambda t_1, \lambda t_2, \lambda c)$, the bivariate Poisson $\wedge$ Gamma has pgf

$$\int_0^\infty \exp \{\lambda(t_1 - c)(z_1 - 1) + \lambda(t_2 - c)(z_2 - 1) + \\ + \lambda c(z_1 z_2 - 1)\} \lambda^{s-1} a^s e^{-\lambda a} d\lambda, \tag{7.36}$$

which is equal to

$$a^s \{a + t_1 + t_2 - c - z_1(t_1 - c) - z_2(t_2 - c) - z_1 z_2 c\}^{-s}. \tag{7.37}$$

Equation (7.37) corresponds to the pgf for model III in Example 7.4. We return to this topic in section 7.8.

Other compounding distributions may of course be used. For example, Bhattacharya and Holla (1965) used the uniform distribution (cf. Exercise 6.15).

7.8 Accident proneness

Continuing the discussion of section 6.6, if it is assumed that

(a) the liability to an accident varies between members of a population, but does not change over time for an individual;

(b) the number of accidents incurred by an individual in a given time-period is Poisson distributed (that is, accidents occur at random);

and (c) the numbers of accidents incurred by an individual in two non-overlapping time-periods are independent;

then a compound Poisson such as (7.37) or (7.38), with $c = 0$, may be used to represent the probability distribution of accidents incurred. Dropping assumption (c) corresponds to the more general case $c \neq 0$.

In the univariate case, corresponding to a single time-period, the accident proneness (apparent contagion) and real contagion models are indistinguishable, as both may lead to the negative binomial distribution. Both assumptions also give rise to a bivariate NBD, but, as Bates and Neyman (1952a, b) pointed out, the two models can now be distinguished. Consider two non-overlapping time-periods, T_1 and T_2, of length t_1 and t_2 respectively. When assumption (c) holds, the simpler case, the bivariate NBD reduces to a negative multinomial with $m = 2$. If the real contagion model has number of occurrences λ per unit time, then the parameters of the models are

	Compound		Generalised	
	T_1 only	T_1 and T_2	T_1 only	T_1 and T_2
Index	s	s	$\lambda a t_1$	$\lambda a (t_1 + t_2)$
Probabilities	$t_1/(a + t_1)$	$t_i/(a + t_1 + t_2)$	q_1	q_i

From the table, the different structure of the parameters is clear. If we consider several time-periods the index for the real contagion model will increase linearly with time, while that for the accident proneness model remains constant. This provides a basis for constructing a test to distinguish the models. Bates (1955) extended the model to k time-periods, also allowing combinations of both real and apparent contagion. When the times between successive events are known for each individual, UMP tests for the presence of linear contagion are available; see Bates' work for details.

Assumption (c) in the Bates—Neyman formulation is sometimes

undesirable, and Edwards and Gurland (1961) explored the use of the correlated compound bivariate Poisson model (7.37). It is also possible to relax assumption (a) somewhat, and suppose that the rate of accidents per unit time is λ in T_1, but $\lambda\gamma$ in T_2, where γ is the same for all individuals. When the numbers of events in the two time-periods are not independent, the question arises — can the results from the first time-period be used to make predictions for the second? The major use of such predictions would be to "weed out" high-risk members of the population. Following Bhattacharya (1967), the conditional density of λ, given that $R_1 = r$, is

$$g^*(\lambda \mid r) = \frac{\lambda^r e^{-\lambda t_1} g(\lambda)}{\int_0^\infty \lambda^r e^{-\lambda t_1} g(\lambda)\, d\lambda}, \qquad (7.38)$$

where $g(\lambda)$ is the compounding density function. For the second time-period the random variable R_2 has conditional expectation (Johnson, 1957)

$$E(R_2 \mid R_1 = r, \lambda) = \lambda t_2 + (t_2 - c)r/t_1,$$

so that

$$E(R_2 \mid R_1 = r) = \int_0^\infty E(R_2 \mid R_1 = r, \lambda)\, g^*(\lambda \mid r)\, d\lambda$$

$$= (t_2 - c)r/t_1 + t_2 \int_0^\infty \lambda g^*(\lambda)\, d\lambda. \qquad (7.39)$$

If only individuals with r_0 or fewer accidents are retained, this selection is helpful in reducing accident rates if

$$E(R_2) > E(R_2 \mid R_1 = r_0). \qquad (7.40)$$

In particular, when $r_0 = 0$, this is true whenever

$$t_2 \int_0^\infty \lambda g(\lambda)\, d\lambda \int_0^\infty e^{-\lambda t_1} g(\lambda)\, d\lambda > (t_2 - c) \int_0^\infty \lambda e^{-\lambda t_1} g(\lambda)\, d\lambda. \qquad (7.41)$$

Condition (7.41) holds whenever the correlation between λ and $\exp(-\lambda t_1)$ is < 0 for $c \geqslant 0$. Since λ is an increasing, and $\exp(-\lambda t_1)$ is a decreasing function of λ, the strict inequality holds for any non-degenerate $g(\lambda)$. The higher the value of c, the more efficient will be the "weeding-out" process. Generally, r_0 should be the largest integer for which (7.40) is satisfied.

7.9 Inference for contagious models

Allowing the accident rates to be λ, $\lambda\gamma$ in time-periods T_1 and T_2 respectively, the Bates–Neyman model has pgf proportional to

$$\{1 - \lambda(z_1 - 1) - \lambda\gamma(z_2 - 1)\}^{-s}. \tag{7.42}$$

The ML estimators for this three-parameter distribution are given by

$$\left.\begin{array}{l} \hat{\lambda}\hat{s} = \bar{r}_1 \\[2mm] \hat{\gamma} = \bar{r}_2/\bar{r}_1 \end{array}\right\} \tag{7.43}$$

and $\hat{s}$ as the solution to

$$\ln\left(1 + \frac{\bar{r}_1 + \bar{r}_2}{s}\right) = \sum_{j=0}^{\infty} \{1 - \text{prob}\,(R_1 + R_2 \leqslant j)\}/(s + j). \tag{7.44}$$

Since the distribution of $R_1 + R_2$ is negative binomial with parameters s and $\lambda(1 + \gamma)/\{1 + \lambda(1 + \gamma)\}$, (7.44) may be solved iteratively in the usual way (see Exercise 5.11).

For the correlated compound model, the pgf is proportional to

$$[1 - \lambda\{(1 - c)(z_1 - 1) + (\gamma - c)(z_2 - 1) + c(z_1 z_2 - 1)\}]^{-s}. \tag{7.45}$$

Equations (7.43) continue to hold, but the expressions for c and s are intractable, as $R_1 + R_2$ no longer follows a NBD (Patil, 1965b). Assuming γ to be known, Edwards and Gurland (1961) derived moment estimators for the remaining parameters. Their results may readily be extended to cover γ unknown, yielding

$$\left.\begin{array}{l} \tilde{\lambda}\tilde{s} = \bar{r}_1, \quad \tilde{\gamma} = \bar{r}_2/\bar{r}_1 \\[2mm] \tilde{s} = (\bar{r}_1)^2\,\{(1 - \theta)(m_{20} - \bar{r}_1)^{-1} + \theta\gamma^2(m_{02} - \bar{r}_2)^{-1}\} \\[2mm] \tilde{c} = m_{11}/\bar{r}_1 - \bar{r}_2/\tilde{s}, \end{array}\right\} \tag{7.46}$$

where m_{ij} denotes the (i, j)th sample moment, i referring to the first time-period. Edwards and Gurland took $\theta = 1$, but this parameter is at choice, and consideration of the asymptotic variance of $\tilde{s}$ suggests that $\theta^{-1} = 1 + \gamma^2$ is a good choice. Substituting this value of θ, with $\tilde{\gamma}$ for γ, yields

$$\tilde{s} = \frac{(\bar{r}_1\bar{r}_2)^2(m_{20} + m_{02} - \bar{r}_1 - \bar{r}_2)}{(\bar{r}_1^2 + \bar{r}_2^2)(m_{20} - \bar{r}_1)(m_{02} - \bar{r}_2)}. \tag{7.47}$$

Subrahmaniam (1966) gives a test of the null hypothesis $H_0 : c = 0$.

7.10 An example

To illustrate the fitting of a bivariate model we reproduce an example given by Mellinger *et al.* (1965). The data form part of a study on accident repeatedness among children in California. Poisson models were fitted to both the marginals and the bivariate distribution using ML estimators, and were rejected using a χ^2 test at extreme significance levels. The estimators (7.46) lead to a negative $\tilde{c}$. The Bates–Neyman model was therefore fitted with ML estimators $\hat{m} = \hat{\lambda}\hat{s} = 1{\cdot}132$, $\hat{\gamma} = 1{\cdot}176$ and $\hat{s} = 3{\cdot}18$. The observed and expected values appear in Table 7.3. The divisions in the table show the groupings used for testing the goodness of fit.

Table 7.3

Observed and expected bivariate distributions of white male children by number of medically attended injuries[1]

(k) No. of injuries in period T_1 (ages 4–7)		(m) Number of injuries in period T_2 (ages 8–11)								No. of children
		0	1	2	3	4	5	6	7 or more	
0	Obs.	101	76	35	15	7	3	3		240
	Exp.	100·3	75·3	37·1	15·1	5·5	(4·4)			
1	Obs.	67	61	32	14	12	4	1	1	192
	Exp.	63·9	63·0	38·5	18·7	7·9	3·1			
2	Obs.	24	36	22	15	6	1	2	1	107
	Exp.	26·7	32·7	23·8	13·5	(9·3)		(3·0)		
3	Obs.	10	19	10	5	2	4		2	52
	Exp.	9·2	13·5	11·4	7·4	(9·9)			(4·2)	
4	Obs.	1	7	3	4	2				17
	Exp.	(4·0)	(7·1)	(7·3)	(4·8)					
5	Obs.	2	1	4	2					9
	Exp.									
6 or more	Obs.	1	1	1	1					4
	Exp.									
No. of children		206	201	107	56	29	12	6	4	621

[1] In "super-cells" one expected value is shown in parentheses for all component cells combined. (Reproduced from Mellinger *et al.* (1965).)

To check whether or not the data exhibit accident proneness, we calculate the moment estimators for the age-groups 4–7, 8–11 and 4–11. These are, approximately,

	4–7	8–11	Combined
k	2·76	8·24	3·60
q	0·32	0·26	0·40

clearly demonstrating accident proneness rather than some form of clustering.

We may conclude that the model fits the data extremely well, and

172

the discussion in the original paper suggests that the theoretical considerations listed in section 7.9 are fairly realistic.

7.11 Bivariate series expansions

If the marginal densities of random variables R_1 and R_2 are $f_1(r_1)$ and $f_2(r_2)$ respectively, then a bivariate density function for R_1 and R_2 when $\rho = \text{corr}(r_1, r_2)$ is given by

$$h(r_1, r_2; \rho) = f_1(r_1)f_2(r_2)\{1 + c_1 p_1(r_1)q_1(r_2) + c_2 p_2(r_1)q_2(r_2) + \ldots\},$$
(7.48)

where p_j and q_j are the jth orthogonal polynomials for f_1 and f_2 respectively. If these polynomials are suitably scaled, we can put $c_1 = \rho\{\text{var}(r_1)\,\text{var}(r_2)\}^{\frac{1}{2}}$ and choose the higher c_j to produce appropriate patterns among the higher order moments, provided they are chosen so as to ensure that h is a bivariate density function. When f_1 and f_2 are both Poisson, the Charlier polynomials defined in (5.51) are used, giving

$$h(r_1, r_2; \rho) = \exp(-\lambda_1 - \lambda_2)\frac{\lambda_1^{r_1}\lambda_2^{r_2}}{r_1!\,r_2!}\{1 + \rho(\lambda_1\lambda_2)^{\frac{1}{2}}p_1 q_1 + \ldots\},$$
(7.49)

and the series may be truncated after an appropriate number of terms. This expression was first obtained by Campbell (1934), who took $c_j = c_1^j/j!$. Aitken and Gonin (1935) derived a similar series for the binomial using the same c_j, while Krishnamoorthy (1951) extended these expansions to cover the multivariate Poisson and binomial forms.

These expansions are particular examples of Lancaster's (1958) general model (see section 3.7), and other densities could be specified in a similar way. Perhaps the most useful application of such series, however, is in situations where the marginal densities are different. For example, (7.48) could be used to represent a joint density function where one variate was discrete and the other continuous; see Exercise 7.9 for an example.

EXERCISES

7.1 Develop arrow diagrams for the multiple hypergeometric, the multiple negative hypergeometric and the Dirichlet–Pascal, showing the various limiting relationships which exist (compare with Fig. 7.1 and Table 7.1.).

7.2 Find the density function of the bivariate binomial ∧ Dirichlet distribution. (Ishii and Hayakawa, 1960)

7.3   For the density function of the bivariate Poisson distribution, derive the recurrence relation

$$r_1\, f(r_1, r_2) \;=\; \lambda_1\, f(r_1 - 1, r_2) + \lambda_{12}\, f(r_1 - 1, r_2 - 1).$$

Hence find a recurrence relation for the p-variate Poisson.

(Teicher, 1954b)

7.4   Find the ML estimators for the negative multinomial with unknown index k and probabilities $\{p_i\}$. (*Hint*: put $\theta_i = p_i/(1 - p_0)$. Cf. Exercise 5.11.)

7.5   For the bivariate Poisson (notation as in Table 7.2) with density function $f(r_1, r_2)$ show that the ML estimators, given a sample of size n, are the solutions to the equations

$$\hat{\lambda}_i = \overline{r}_i - \hat{\lambda}_{12}, \quad i = 1, 2$$

and

$$\sum \left\{ \frac{f(r_1 - 1, r_2 - 1)}{f(r_1, r_2)} \right\} = n,$$

where the summation and means are taken over the whole sample.

(Holgate, 1964)

7.6   Following the notation of section 7.6, let $\mu'(\mathbf{s})$ denote the non-central moment of order $\mathbf{s}$ of a MGPSD. Show that

$$\mu'(\mathbf{s} + \mathbf{l}_j) = \theta_j \frac{\partial \mu'(\mathbf{s})}{\partial \theta_j} + \mu'(\mathbf{l}_j)\, \mu'(\mathbf{s}). \quad \text{(Patil, 1965a)}$$

7.7   For the multiple LSD given in Table 7.2, show that the regression function for r_1 on $r_2, \ldots, r_m$ is

$$E(r_1 \mid r_2, \ldots, r_m) = q_1/\{(1 - q_1) \ln(1 - q_1)\}, \quad R = 0$$
$$= q_1 R/(1 - q_1), \quad R > 0$$

where $R = \sum_{j=2}^{m} r_j.$ (Patil and Bildikar, 1967)

7.8   Apply the selection rule given in (7.40) to find the value of r_0 when model (7.37) holds.

7.9   Using expansion (7.48), write down a joint density function for a Poisson and a normal variate. Hence evaluate the conditional distributions and show that the regression functions are both linear.

CHAPTER 8

APPROXIMATIONS AND VARIANCE STABILISING TRANSFORMS

8.1 Introduction

The subject-matter of this chapter is somewhat different in nature from the rest of the monograph, but is judged of sufficient importance to warrant inclusion. The development of approximations to distribution functions might be said to have started with De Moivre's (1730) derivation of the normal limit for the binomial, and a substantial body of literature has accumulated since that time. It would be impossible to do justice to more than a tiny fraction of this work here, and a major discussion is not attempted. After a brief review, in section 2, of some general methods, we concentrate in sections 3 to 7 on methods which are useful for the distributions discussed earlier in this monograph. Particular variance-stabilising transforms are considered in section 8, and a link between approximation methods and variance-stabilising transforms is forged in section 9.

8.2 General approximation methods

The mathematical development of asymptotic expansions has been presented by several authors, Erdelyi (1956) and De Bruijn (1961), for example; while an excellent review of the state of the art with reference to probability distributions appears in Wallace (1958). Approximations for discrete distributions are fully discussed in a new monograph by Molenaar (1970), while Mayne (1972) provides a comprehensive review for both discrete and continuous distributions. These works, therefore, allow our treatment to be of a brief and largely prescriptive nature.

The general problem may be formulated in the following way:

If the random variable X has distribution function $F(x) = $ prob $(X = x)$, we may wish to approximate either (a) $F(x)$ given x, or (b) x given $F(x)$, the inverse form. Our discussion centres mainly on (a), but the importance of (b), particularly for hypothesis testing and interval estimation, must not be overlooked.

We are mainly concerned, therefore, with the evaluation of known distribution functions at one or more values of the random variable, and we shall largely ignore methods of interpolation and also approximations to sampling distributions. Of course, the distributions considered in this monograph may be used to approximate sampling distributions, so that the problems are not unconnected. The approaches given below are described with reference to continuous distributions, but could be used for discrete types also.

8.3 Continued fractions and Mills' ratio

The expression

$$\frac{a_0}{b_0 +} \frac{a_1}{b_1 +} \frac{a_2}{b_2 +} \cdots \frac{a_n}{b_n +} \cdots \tag{8.1}$$

is referred to as a "continued fraction", and is sometimes written as

$$\cfrac{a_0}{b_0 + \cfrac{a_1}{b_1 + \cfrac{a_2}{\ddots \; + \cfrac{a_n}{b_n + a_{n+1}} \ddots}}}$$

with the values

$$\frac{a_0}{a_1}, \quad \frac{a_0 b_1}{b_0 b_1 + a_1}, \quad \frac{a_0 (b_1 b_2 + a_2)}{a_1 b_2 + b_0 (b_1 b_2 + a_2)}, \cdots$$

when the fraction is terminated after one, two, three, ... terms. Continued fractions have been most used in conjunction with Mills' ratio, written as

$$M(x) = \{1 - F(x)\}/f(x), \tag{8.2}$$

the ratio of the "tail area" of the curve to its bounding ordinate. When $f(x)$ can be written in closed form, but $F(x)$ cannot (as with the normal), an approximation for $M(x)$ allows $F(x)$ to be evaluated. Approximations to $M(x)$ are usually easier to derive than those for $F(x)$.

Example 8.1

For the standard normal curve,

$$M(x) = \exp\left(\tfrac{1}{2} x^2\right) \int_x^\infty \exp\left(-\tfrac{1}{2} u^2\right) du, \quad x > 0$$

$$= (2\pi)^{-\frac{1}{2}} \int_{-\infty}^\infty \exp\left(-\tfrac{1}{2} x^2 u^2\right)(1 + u^2)^{-1} du$$

$$= (x/2\sqrt{\pi}) \int_{-\infty}^\infty e^{-v} \{v(v + \tfrac{1}{2} x^2)\}^{\frac{1}{2}} dv.$$

See, for example, Kendall and Stuart (1969, pp. 136–7) for a detailed justification of these steps. Expanding in terms of $\frac{1}{2}x^2$, and integrating term by term, we obtain Laplace's expansion

$$M(x) = \frac{1}{x} - \frac{1}{x^3} + \frac{1.3}{x^5} - \ldots (-1)^j \frac{1.3.5\ldots(2j-1)}{x^{2j+1}} + Mj. \quad (8.3)$$

Although the series ultimately diverges for any x, the remainder term M_j is less in absolute value than

$$x^{-2j-1} 1.3.5. \ldots (2j-1).$$

A suitable truncation point may be obtained by choosing j to minimise the absolute error, given x; thus, use j terms when $2j - 1 \leqslant x^2 \leqslant 2j + 1$. However, the accuracy of the approximation is much improved if expansion (8.3) is reformulated as a continued fraction. Using Viskovatoff's method (Khovanskii, 1963, p. 27), we find that

$$M(x) = \frac{1}{x+} \frac{1}{x+} \frac{2}{x+} \ldots \frac{n}{x+}.$$

Likewise, for small $x > 0$, we may use the Mills' ratio $\bar{M}(x) = \{F(x) - F(0)\}/f(x)$ which yields for the normal curve the continued fraction (Shenton, 1954)

$$\bar{M}(x) = \frac{x}{1-} \frac{x}{3+} \frac{2x}{5-} \frac{3x}{7+} \ldots . \quad (8.4)$$

Truncation of (8.3) and (8.4) after j terms gives rational approximations to $F(x)$ in terms of polynomials in x. For a discussion of rational approximations, see Hastings (1957). Other approaches for the normal are summarised in Table 8.1, pages 182–3.

8.4 Series expansions

Systems such as the Pearson curves may be used to provide approximations when only the first four moments are known (Moore, 1957; E.S. Pearson, 1963), and the same applies to the Gram–Charlier, Edgeworth and Cornish–Fisher expansions discussed in Chapter 2. In particular, Esseen (1945) gives a revised form of the Cornish–Fisher expansion for use with lattice type discrete random variables (cf. section 5.13). A further series of value is the Hotelling–Frankel expansion, which we now consider. Hotelling and Frankel (1938) showed, given the density function $f(x, \theta)$, where f may depend on other parameters besides θ, that it is possible to represent f as

$$f(x, \theta) = h(x) \{1 + \sum_{j=1}^{\infty} h_j(x) \theta^{-j}\}, \quad (8.5)$$

where $\lim_{\theta \to \infty} f(x, \theta) = h(x)$, and the $h_j(x)$ are functions of x and the remaining parameters only. Under suitable extensive conditions, the convergence of the infinite series, when f and h are both continuous, was proved by Wasow (1956). This proof can be extended to cover (i) f and h both discrete, and (ii) f discrete, h continuous, if $f(r, \theta) =$ prob $(R = r)$ is taken as prob $(r - \frac{1}{2} \leqslant R < r + \frac{1}{2})$ in such a way that the resulting density function is continuous.

However, interest will usually focus on the truncated series, when we approximate $F(x, \theta) = \int_{-\infty}^{x} f(x, \theta) \, dx$ by

$$H(x) + \sum_{j=1}^{k} \theta^{-j} \int_{-\infty}^{x} h(u) \, h_j(u) \, du, \tag{8.6}$$

for a suitable choice of k.

If, instead of expressing one distribution function in terms of another, we wish to put

$$F(t, \theta) = H(x) \tag{8.7}$$

and find an expression for x in terms of t, we obtain

$$f(t, \theta) = h(x)(dx/dt). \tag{8.8}$$

Putting $x = \sum_{j} a_j \theta^{-j}$, where the $\{a_j\}$ are functions of t, and substituting (8.5) into (8.8), we obtain series in powers of θ^{-1}. Comparing coefficients of terms in θ^{-j}, we obtain a differential equation in a_j which may be solved for a_j successively, $j = 1, 2, \dots$.

From the general form of (8.6), it is clear that this expansion involves equating all the finite moments of F and H up to and including terms of order θ^{-k}. This generally produces more accurate approximations than equating the first few moments and ignoring the remainder (Ord, 1968a).

Example 8.2 (Hotelling and Frankel, 1938)

Let $h(x)$ denote the standard normal density function and $f(t, m)$ the Student's t (Type VII) density with $m \equiv \theta$ degrees of freedom, that is,

$$f(t, m) = \frac{\Gamma(\frac{1}{2}m + \frac{1}{2})}{(m\pi)^{\frac{1}{2}} \Gamma(m)} (1 + t^2/m)^{-\frac{1}{2}(m+1)}.$$

Using Stirling's approximation, it follows that

178

$$f(t, m) = (2\pi)^{-\frac{1}{2}} \exp\left\{-\tfrac{1}{2}t^2 + \frac{1}{4m}(t^4 + 2t^2 - 1) - \frac{1}{12m^2}(2t^6 - 3t^4)\right.$$
$$\left. + O(m^{-3})\right\}, \tag{8.9}$$

from which an expansion of form (8.5) follows. An expansion such as (8.6) for $F(t, m)$ is given by integration with respect to t. Using (8.8), the normal variable corresponding to t is

$$x = t\left\{1 - \frac{1}{4m}(t^2 + 1) + \frac{1}{12m^2}(13t^4 + 8t^2 + 3) + O(m^{-3})\right\}, \tag{8.10}$$

a useful approximation formula, provided that $m > t^2$.

Example 8.3 (Ord, 1968a)

Let $h(r) = h(r \mid n)$ denote the binomial with parameters n and p, and $f(r, N)$ the hypergeometric with parameters n, p and $N \equiv \theta$. Thus

$$f(r, N) = \binom{Np}{r}\binom{Nq}{n-r}\Big/\binom{N}{n}, \quad r = 0, 1, \ldots,$$

yielding, after some algebra, the cumulative distribution

$$F(r, N) = \sum_{j=0}^{r} f(r, N)$$
$$= H(r\mid n) - \frac{n(n-1)p}{2Nq}\{H(r\mid n) - 2H(r-1\mid n-1) + H(r-2\mid n-2)\} +$$
$$+ O(N^{-2}), \tag{8.11}$$

where $H(r \mid n) = \sum_{j=0}^{r} h(j \mid n) = \sum_{j=0}^{r}\binom{n}{j}p^j q^{n-j}$;

$H(r \mid n)$ may be evaluated using tables of the beta distribution; see equation (8.15).

Approximation (8.11) is reasonably accurate, provided that r or $s = n - r$ is used, according to whether $p < \tfrac{1}{2}$ or $p > \tfrac{1}{2}$. However, it may be asked whether the approximation could be improved by reparametrising the distribution. In the example, if we take

$$N_0 = N - a,$$
$$p_0 = (Np - b)/N_0,$$

it can be shown that the term of order N_0^{-1} has a zero coefficient if three simple conditions hold. These conditions are linearly dependent and reduce to $2a = n - 1$, $2b = r$, leading to the improved approximation

$$F(r, N) = H(r \mid n) + \frac{n(n-1)}{24N_0^2} \{(n+1)H(r \mid n) - (n-r+1)H(r \mid n-2) -$$

$$- (r+2)H(r-2 \mid n-2) + 2H(r-1 \mid n-2)\} + O(N_0^{-4}).$$

(8.12)

This approximation was first derived by Wise (1954), using a different argument; even the first term is remarkably accurate, and the first two terms are adequate for almost all practical purposes.

The argument need not stop here, however, and one could employ series expansions for the parameters, derived from (8.5) as before. Since, for $h(r)$ discrete, a useful series of type (8.10) cannot be found, the parameter series approach is often attractive, as the parameters of interest often vary continuously, or are sufficiently large to be approximated by a continuous variable. This approach has been used to approximate the binomial with the Poisson by Wise (1950), Bolshev *et al.* (1961) and Molenaar (1970) — see Table 8.2 (pages 184–5). The general idea seems capable, and worthy, of further exploitation.

The methods of this section can be extended in various ways, to cover f discrete, h continuous, for example (Exercise 8.3). When a distribution has several parameters, and an approximation in terms of a standard from, such as the normal, is required, one could either carry out a single step like (8.5) or else a series of steps, e.g.

hypergeometric → binomial → Poisson → normal →
approximation to normal cdf;

one or more steps could be omitted. The advantage of this second approach is that the various ways in which the limiting forms arise can be taken into account to select the most suitable approximation (e.g. $N \to \infty$, $0 < p < 1$ and n finite *or* $N \to \infty$, $p \to 0$ and $0 < n/N < 1$ for the hypergeometric → binomial link). The form of limiting process may also affect the choice of approximating distribution (binomial to Poisson or normal, for example).

8.5 Accuracy and the construction of approximations

So far we have said very little about the accuracy of different approximations or about methods of assessing the accuracy. One approach is to develop limit theorems and obtain the order of magnitude of the error (e.g. in terms of powers of θ^{-1} as above). For example,

Esseen (1945) and Gnedenko and Kolmogorov (1954) have made major contributions in this area. The paper by Govindarajulu (1965) also covers this topic, as well as reviewing the progress made in evaluating absolute bounds for the error when discrete distributions are approximated using the normal. Regrettably, this major field of inquiry cannot be further considered here.

Various measures have been employed in empirical studies to assess accuracy of approximation, $H(r)$, to the distribution function $F(r)$, the relative tail error,

$$
\begin{aligned}
E_r &= 100 \cdot \frac{H(r) - F(r)}{F(r)}, \quad F(r) \leqslant 0{\cdot}5, \\
&= 100 \cdot \frac{F(r) - H(r)}{1 - F(r)}, \quad \text{otherwise,}
\end{aligned}
\right\}
\tag{8.13}
$$

being perhaps the most suitable. Summary statistics such as $\max\limits_{r} E_r$ can also be used. The studies in Molenaar (1970) for the Poisson and binomial are good examples of this empirical approach.

Trying to prescribe an approximation for general use is rather like administering a patent medicine — as long as no serious harm is done, one's conscience is reasonably clear. The summary tables presented try to indicate "how, which and when", with associated references to explain "why". The emphasis is on methods which are relatively easy to apply, although standard tables and interpolation methods are a general requirement. Most approximations given are for the cumulative distribution function, although these forms can sometimes be inverted to give the quantiles (see Exercise 8.4 for an example).

It should be recalled in passing that certain simple relationships between cdf's exist, reducing the number of approximations and tables required. The Poisson and the gamma are related, since

$$
\sum_{j=0}^{r} \lambda^j e^{-\lambda}/j! = \frac{1}{\Gamma(r+1)} \int_{\lambda}^{\infty} x^r e^{-x}\,dx,
\tag{8.14}
$$

while if Y is a Pearson Type V variate, $X = Y^{-1}$ is Type III (gamma). Likewise the beta and the binomial obey the relation

$$
\sum_{j=0}^{r} \binom{n}{j} p^j (1-p)^{n-j} = \frac{1}{\beta(r+1,\, n-r)} \int_{p}^{1} u^r (1-u)^{n-r-1}\,du,
\tag{8.15}
$$

while the negative binomial and F link up to provide

$$\sum_{j=0}^{r} \binom{k+j-1}{j} w^j(1+w)^{-k-j} = \frac{1}{\beta(r+1,\,k)} \int_{w}^{\infty} u^j(1+u)^{-k-j-1} du$$

$$= \frac{1}{\beta(r+1,\,k)} \int_{q}^{1} v^j(1-v)^{k-1} dv, \qquad (8.16)$$

where $q = w(1+w)^{-1}$.

Table 8.1 gives approximations for the normal cdf and quantiles, followed by approximations to the beta and gamma forms in terms of the normal. A list of relevant tables is given in Appendix B.

Exact evaluation of discrete probabilities could be carried out using the recurrence formulae given in earlier chapters, but where the range of the variate is large enough to make this too tedious, approximations based on either continuous or discrete distributions could be used. Generally, inferior approximations have been dropped from the list, but their omission is sometimes mentioned. Approximations which derive from variance-stabilising methods are omitted, and are listed later in the chapter (Table 8.3). Table 8.1 should also be consulted in the light of (8.15) and (8.16).

A few comments on Table 8.2 follow.

Binomial. A very large number of possibilities exist, and many of those not mentioned here are reviewed in Molenaar (1970). The Poisson form suggested in the table seems more accurate than earlier forms given by Wise (1950 – see Table 8.1) and Bolshev *et al.* (1961). Molenaar's empirical results also show the Gram–Charlier Type B expansion (Raff, 1956) to be somewhat less efficient. Recent work by Borges (1970) deserves mention as his integral transform method reduces the error to $O(n^{-1})$ for all p, but this approach is rather cumbersome, and the work by Gebhardt (1969) suggests that the Camp–Paulson form is equally efficient.

Other hypergeometric series distributions. Suggestions are given in Ord (1968a), following the method of Example 8.2. These could be improved using the methods suggested in that example.

8.6 The contagious distributions

Very few satisfactory approximations have been developed for the contagious distributions (except for the negative binomial, covered indirectly above). However, Martin and Katti (1962) demonstrated, for

Table 8.1   *Approximations to*

Distribution[(*)]	Approximation
Normal, $\Phi(x)$	Mills' ratio: $M_1(x) = 1/(x + ae^{-bx})$, $M_2(x) = a/\{(x^2 + b)^{\frac{1}{2}} + cx\}$.
Normal quantiles	Expansions using Chebyshev polynomials. $$K(t) = t - \frac{(2 \cdot 307\ 53 + 0 \cdot 270\ 61t)}{1 + 0 \cdot 992\ 29t + 0 \cdot 044\ 81t^2},$$ or $$K(t) = t - \frac{(2 \cdot 515\ 517 + 0 \cdot 802\ 853t + 0 \cdot 010\ 328t^2)}{1 + 1 \cdot 432\ 788t + 0 \cdot 189\ 269t^2 + 0 \cdot 001\ 308t^3},$$ where $t = \{-2 \ln(1 - F)\}^{\frac{1}{2}}$, $F \geqslant \frac{1}{2}$.
Gamma, $I(x \mid n)$	Treat z as a standard normal deviate, where x is χ^2 with n degrees of freedom. $$z = b\left\{\left(\frac{x - c}{g}\right)^h - a\right\}.$$ Take $I(x \mid n) = \{2\Phi(y) - 1\}^{2n}$, where $y = (hx)^{\frac{1}{2}}$, $h = (\pi/2)\{\Gamma(n + 1)\}^{1/an}$.
Beta, $B(x \mid a, b)$	Treat z as a standard normal deviate, where $$z = 3(ab)^{\frac{1}{2}}\frac{[\alpha\{1 - (9b)^{-1}\} - \beta\{1 - (9a)^{-1}\}]}{a\beta^2 + b\alpha^2},$$ $\beta = (bx)^{1/3}$, $\alpha = \{a(1 - x)\}^{1/3}$. The cdf in terms of the gamma cdf is $1 - B(x \mid a, b) = I(y \mid b) - p(y)/24c^2 + \ldots$, where $p(y) = y^b e^{-y}(b + 1 + y)/\Gamma(b - 1)$, $c = a + \frac{1}{2}b + \frac{1}{2}$ and $y = -c \ln x$.

[(*)] Notation for cdf appears after name.

continuous distributions

Comments and References

$a = 0.8$, $b = 0.4$; good for small x (Hart, 1957).

$a = 4$, $b = 8$, $c = 3$; close upper bound for large x (Sampford, 1953).

$a = \pi$, $b = 2\pi$, $c = \pi - 1$; close lower bound (Boyd, 1959).

$a = \pi/(\pi - 2)$, $b = 2\pi/(\pi - 2)^2$, $c = 2/(\pi - 2)$; close upper bound (Boyd, 1959).

Clenshaw (1962), Ruben (1964).

Hastings (1957), pp. 191–2.

n is the index parameter.

$c = 0$, $g = n$, $h = \frac{1}{3}$, $a = 1 - (2/9n)$, $b = (9n/2)^{\frac{1}{2}}$ (Wilson and Hilferty, 1931).

$c = n/13$, $g = 12n/13$, $h = 5/13$, $a = 1 - (5/18n)\{1 + (7/48n)\}$,
$b = 1.2\{1 - (18n)^{-1}\}(2n)^{\frac{1}{2}}$ (Haldane, 1937).

Accurate except in extreme tails for n not too small. The Haldane form is generally more accurate (cf. section 2.11).

For small n (Pinkham, 1962).

For a general discussion see Teichroew (1957); for Pearson's Type V with index m, take gamma forms with $y = 1/x$ and $n = m - 2$. See also the Poisson approximations in Table 8.2.

a and b are parameters.

Paulson (1942).

Obtained using the methods of section 8.3 (Wise, 1950).
Valid for large a.

See also the binomial approximations in Table 8.2.
For F (Type VI) use $y = x/(1 - x)$, and for Student's t (Type VII), $t = y^{\frac{1}{2}}$.

Table 8.2 *Approximations to*

Distribution	Approximation
Poisson (λ)	z as a standard normal deviate, where $z_1 = 2(r + a)^{\frac{1}{2}} - 2\lambda^{\frac{1}{2}}$; $z_2 = 2\{r + (w^2 + 5)/9\}^{\frac{1}{2}} - 2\{\lambda + (w^2 - 4)/36\}^{\frac{1}{2}}$, where $w = (r + \frac{1}{2} - \lambda)\lambda^{-\frac{1}{2}}$; $z_3 = \left(r + \frac{2}{3} - \lambda + \dfrac{0\cdot02}{r + 1}\right)\{1 + A(v)\}^{\frac{1}{2}}\lambda^{-\frac{1}{2}}$, where $v = (r + \frac{1}{2})/\lambda$.
Binomial (n, p)	z as a standard normal deviate, where $z_1 = -u_1/3u_2$, where $u_1 = c\{9 - (n - r)^{-1}\} + (r + 1)^{-1} - 9$, $\quad\quad u_2^2 = c^2(n - r)^{-1} + (r + 1)^{-1}$, and $\quad c^3 = (n - r)p/(r + 1)q$; $z_2 = B\{1 + A(v) + A(w)\}^{\frac{1}{2}}\{(n + \frac{1}{6})pq\}^{-\frac{1}{2}}$, where $v = (r + \frac{1}{2})/np$, $\quad w = (n - r - \frac{1}{2})/nq$. R as a Poisson variate with parameter λ, where $\lambda_1 = (2n - r)p/(2 - p)$, $\lambda_2 = np\{(12 - 2p)n^2 - 7rn\}/\{(12 - 8p)n^2 - rn + r\}$.
Hypergeometric (N, n, p)	z as a standard normal deviate, where $z = \{r + \frac{1}{2} - (n + 1)\rho\}/\sigma$, where $\rho = (Np + 1)/(N + 1)$, $(N + 2)\sigma = (n + 1)\rho(1 - \rho)(N - n + 1)$. R as a binomial variate with parameters n, p_0, where $p_0 = (Np - \frac{1}{2}r)/(N - \frac{1}{2}n + \frac{1}{2})$. R as a Poisson variate, with parameter $\lambda = \mu + (\mu - r)(2rN - nN + 10rn)/3N^2$, where $\mu = np$.

Note: $A(v) = (1 - v^2 + 2v \ln v)(1 - v)^{-2}$.

Comments and References

$a = 1$ for tail regions, but $a = \frac{3}{4}$ for $0 \cdot 06 \leqslant P \leqslant 0 \cdot 94$ (Molenaar, 1970).

For more accurate work (Molenaar, 1970).

Peizer and Pratt (1968), Pratt (1968). The joint paper gives a table for A in terms of v — see footnote below.

Camp (1951), Paulson (1942).

Peizer and Pratt (1968).

Bolshev (1961).

Molenaar (1970) for more accurate work.

Nicholson (1956).

Wise (1954), more accurate than Burr (1952) and Sandiford (1960) — see Example 8.2, Exercises 8.1 and 8.2.

Molenaar (1970).

the Neyman Type A with parameters λ_1 and λ_2, that

(a) for λ_1 large, $z = (r - \lambda_1\lambda_2)/\{\lambda_1\lambda_2(1 + \lambda_2)\}^{\frac{1}{2}}$ is a standard normal deviate to $O(\lambda_1^{-\frac{1}{2}})$;

(b) for λ_2 small, the random variable R is Poisson distributed with mean $\lambda_1\lambda_2$, to $O(\lambda_2)$;

(c) for λ_1 small, R represents the finite mixture

$$\lambda_1\{\text{Poisson with parameter } \lambda_2\} + (1 - \lambda_1)\delta(r),$$

where $\delta(r) = 1, \quad$ if $r = 0,$
$\qquad\qquad = 0$ otherwise.

These three forms of limiting relationship are typical of the generalised Poisson distributions (see Exercise 8.6).

Douglas (1965) used the steepest descent method to obtain the following approximations:

(a) Neyman Type A (parameters λ_1, λ_2, see Table 6.1)

$$f_r \sim \frac{\lambda_2^r \exp(-\lambda_1 + rz^{-1})}{z^r\{2\pi r(1 + z)\}^{\frac{1}{2}}}, \quad r = 1, 2, \dots,$$

where z is the unique positive root of $\mu z e^z = r$, $\mu = \lambda_1 e^{-\lambda_2}$ and $f_0 = \exp(\mu - \lambda_1)$. In general, Σf_r is bounded but does not equal 1. The approximation is reasonable for relatively large values of r.

(b) Pólya–Aeppli (parameters λ, τ, see Table 6.1)

$$f_r \sim \frac{\tau^r \phi^{\frac{1}{4}} \exp\{\frac{1}{2}(\phi - \tau) + 2(\phi r)^{\frac{1}{2}}\}}{2\pi^{\frac{1}{2}} r^{\frac{3}{4}}},$$

where $\phi = \lambda(1 - \tau)$. This approximation has modes at the same values of r as the original distribution and appears to work well in the upper tail, although further numerical evidence is required before the accuracy of the approximation can be judged.

The generalised power series distributions with parameter θ, defined in section 6.2, are shown (Govindarajulu, 1965) to approach normality if they are unimodal, and the variance approaches infinity as θ becomes large.

8.7 Multivariate approximations

The discussion of multivariate models in Chapter 3 revealed that transformations of the marginal densities may impose strong conditions on the structure of the joint distribution. In such circumstances, the series expansions considered there would seem to be the best approach to pursue. For recent work in this direction, using generalised Edgeworth expansions, see Chambers (1967).

8.8 Variance stabilisation

One of the basic assumptions in analysis of variance models is that the observations come from populations with equal error variances, allowing estimation of the residual variance component. In particular, this implies that the variance does not depend on the mean. When the observations are frequency counts this assumption is invalid (mean = variance for the Poisson, for example). This has led to the use of some function, $u(r)$ say, selected so that var (u) is constant, at least asymptotically; u is then called a "variance-stabilising transform", and the analysis is carried out on the transformed variables. Such a procedure is not without its critics; Fisher (1953), for example, pointed out that interpretation of the results may be difficult, as the model becomes additive in u instead of in r. For this reason, a logarithmic transform which allows of interpretation as a multiplicative model may be preferred to some more esoteric function which gives a more stable asymptotic variance. Bartlett (1947) gives a very sound discussion of the advantages and drawbacks of the method. Box and Cox (1964) have pointed to the need for transformations with a basically simple structure, which stabilise variance and also induce normality. We cannot expect any particular transform to meet all these requirements, although section 8.9 shows that we may be fortunate in some cases. It should be pointed out that Box and Cox were discussing general transforms of the data, whereas we are considering data drawn from a particular type of population.

Hinz and Gurland (1968) show how discrete data may be analysed without resort to transformations, as a development of their minimum chi-square estimation technique discussed in section 6.11. For other methods of analysis, see the monograph by Cox (1970).

For a given distribution with density function $f(r, \theta)$, θ a parameter, let us consider the transformation $u \equiv u(r)$. A suitable form for u, given f, can be obtained by the method of Tippett (1934) and Beall (1942). Expanding u in a Taylor series about the mean m, we have

$$u(r) = u(m) + (r - m)u'(m) + (r - m)^2 u''(m)/2 + \dots . \qquad (8.17)$$

If $u^{(j)}(m)$ is $O(\theta^{(1-2j)\delta})$ and $E\{(r-m)^j\}$ is $O(\theta^{j\delta})$, $\delta > 0$, then for large θ,

$$
\begin{aligned}
\mathrm{var}\,(u) &= E[\{u - u(m)\}^2] \\
&= \{u'(m)\}^2\,\mathrm{var}\,(r)\{1 + O(\theta^{-\delta})\}.
\end{aligned}
\tag{8.18}
$$

The conditions given are sufficient, but may be relaxed in specific instances. From (8.18), if var (u) is to be a constant, u is the solution to the differential equation

$$
u'(m) \propto \{\mathrm{var}\,(r)\}^{-\frac{1}{2}}.
\tag{8.19}
$$

When R is Poisson distributed, for example, var $(r) = m$, leading to $u(r) = 2r^{\frac{1}{2}}$ if u is scaled to have variance 1 ($\theta^2 = m$ and $\delta = \frac{1}{2}$ in this case, though (8.18) is correct to order -2δ). Equation (8.19) forms the basis of construction of most variance-stabilising transforms, although the u functions are often modified to take account of higher order terms in (8.17), increasing the degree of variance stability.

Example 8.4 (Ord, 1967a)

For the random variable R with density function $f(r, n)$ and finite moments $m'_1 = m$, m_i, $i \geqslant 2$, consider the transform

$$
u(r) = c(n + d)^{\frac{1}{2}}\,\arcsin\left(\frac{r + a}{n + b}\right)^{\frac{1}{2}}.
$$

From (8.17) we can obtain a series expansion for $u(r)$ about m, leading to

$$
\begin{aligned}
\mathrm{var}\,(u) = \frac{c^2(n + d)}{4}&[m_2\phi - \tfrac{1}{2}m_3(n + b - 2m - 2a)\phi^2 + \\
&+ \frac{1}{48}\phi^3\{4m_4(8m^2 - 8mn + 3n^2) + 3(m_4 - m_2^2)(2m + 2a - n - b)^2\} + \\
&+ O(n^{-2})],
\end{aligned}
$$

where $\phi = (m + a)(n + b - m - a)$.

For the binomial, $c = 2$, $b = 2a = \frac{3}{4}$ and $d = \frac{1}{2}$ give var $(u) = 1 + O(n^{-2})$, while for the hypergeometric with parameters N, n and p, the coefficients

$$
c = 2\left(\frac{N - 1}{N - n}\right)^{\frac{1}{2}}, \quad b = 2a = \tfrac{3}{4} + n/8N \quad \text{and} \quad d = \tfrac{1}{2} + 3n/2N
$$

give var $(u) = 1 + O(n^{-2})$.

Various transforms have been proposed for the classical discrete distributions and these are summarised in Table 8.3; tables of these transforms, where available, are listed in Appendix B.

As Cochran (1940, 1943) had demonstrated, the treatment of zero frequencies is most important, explaining the value of transforms with non-zero a in Table 8.3. Empirical studies by Eisenhart (1947), Anscombe (1948) and Laubscher (1961) suggest that the Anscombe transform is best for the binomial when the probability of success, p, lies in the range $0\cdot1 < p < 0\cdot9$, and the Laubscher form otherwise. Similar conclusions are plausible for the negative binomial, but there seems little to choose between the two more accurate Poisson transforms. The logarithmic transform does not perform very well in comparison with the other alternatives. For the hypergeometric only a very small amount of work has been done (Ord, 1967a), but the conclusions parallel those drawn for the binomial. Similar results may be anticipated for other hypergeometric series distributions.

8.9 Transforms and approximations

Several of the variance-stabilising transforms discussed in the last section have been recommended as approximations, treating u as a normal variate. Blom (1954), for example, recommends the arcsine transform for the binomial. To see why this should be so, we extend an argument of Jenkins and Watts (1969, p. 125).

The log-likelihood, l, of the normal distribution is quadratic, both in the random variable and its location parameter μ. Thus

$$\frac{d^2 l}{d\mu^2} = -c^2, \text{ a constant.} \qquad (8.20)$$

If the transformation $\mu = u(\theta)$ is carried out on the location parameter, θ, of a non-normal distribution, so that (8.20) holds, we have

$$\frac{d^2 l}{d\mu^2} = \left(\frac{d^2 l}{d\theta^2}\right)\left(\frac{d\theta}{d\mu}\right)^2 + \frac{dl}{d\theta} \cdot \frac{d^2\theta}{d\mu^2} = -c^2, \qquad (8.21)$$

which could be solved to obtain μ. In general, (8.21) cannot be solved analytically, but if we consider the value of the equation at $\hat{\theta}$, the maximum likelihood estimator for θ,

$$\frac{dl}{d\theta} = 0 = \left(\frac{dl}{d\mu}\right)\Big/\left(\frac{d\theta}{d\mu}\right) \quad \text{at } \theta = \hat{\theta},$$

Table 8.3 *Variance stabilising transforms*

Distribution	Transformation	References and comments
Poisson	$P(a) = (r + a)^{\frac{1}{2}}$	Bartlett (1936); stable to $O(n^{-1})$.
	$2P(0)$	
	$P(0) + P(1)$	Freeman and Tukey (1950), also to $O(n^{-1})$, but better for λ near zero.
	$2P(\frac{3}{8})$	Anscombe (1948) $\Big\}$ to $O(n^{-2})$
	$P(0) + P(\frac{3}{4})$	Laubscher (1960)
Binomial	$B(a, b, d) = (n + d)^{\frac{1}{2}} \arcsin \left(\dfrac{r + a}{n + b} \right)^{\frac{1}{2}}$	
	$2B(0, 0, 0)$	Bartlett
	$B(0, 1, 1) + B(1, 1, 1)$	Freeman–Tukey
	$2B(\frac{3}{8}, \frac{3}{4}, \frac{1}{2})$	Anscombe
	$B(0, 0, 0) + B(\frac{3}{4}, \frac{3}{2}, 1)$	Laubscher (1961)
Negative binomial	$N(a, b, d) = (k - d)^{\frac{1}{2}} \operatorname{arcsinh} \left(\dfrac{r + a}{k + b} \right)^{\frac{1}{2}}$	
	$2N(0, 0, 0)$	Beall (1942)
	$2N(\frac{3}{8}, \frac{3}{4}, \frac{1}{2})$	Anscombe
	$N(0, 0, 0) + N(\frac{3}{4}, \frac{3}{2}, 1)$	Laubscher (1961)
	$\ln (r + \frac{1}{2}k)$	Bartlett and Kendall (1946)

Notes: (1) All transformations, except the logarithmic, are scaled to have asymptotic variance 1·0.

(2) Order of accuracy of transform for binomial and negative binomial forms as for the corresponding Poisson form.

leading to

$$\frac{d^2l}{d\mu^2} = \left(\frac{d^2l}{d\theta^2}\right)_{\hat{\theta}} \left(\frac{d\theta}{d\mu}\right)^2 = -c^2,$$

so that

$$\mu = u(\theta) = c \int \left\{\frac{-d^2l}{d\theta^2}\right\}^{\frac{1}{2}} d\theta, \qquad (8.22)$$

for $\theta = \hat{\theta}$.

On this basis, transform (8.22) can be recommended as a normalising transform, but since var $(\hat{\theta}) = -1/E(d^2l/d\theta^2)$ asymptotically, $u(\hat{\theta})$ will have constant variance. Therefore, whenever $\hat{\theta} = r$ for a single observation, (8.19) and (8.22) will lead to the same transformation; the power series distributions clearly fall into this category as $\hat{\theta} = r$ is a defining property of the exponential/GPSD family.

Example 8.5 (Jenkins and Watts, 1969)

For the binomial, $\theta = p$ and

$$\frac{d^2l}{dp^2} = \frac{-n}{p(1-p)},$$

leading to $u(p) = c \int \{p(1-p)\}^{-\frac{1}{2}} dp = c \arcsin (p^{\frac{1}{2}})$.

A more general problem relating to the use of normalising transforms is the estimation of parameters relating to the original random variable. Let the random variable X be transformed to $Y = g(X)$. If $t = t(Y)$ is the minimum-variance unbiased estimator (MVUE) for $\mu = E(Y)$, $g^{-1}(t)$ will not, in general, be the MVUE for $\theta = E(X)$. The evaluation of $\tilde{\theta}$, the MVUE for θ, in terms of t was first explored by Finney (1941) for the logarithmic transform, while Neyman and Scott (1960) derived a general expansion in t for $\tilde{\theta}$ using Taylor's series. Hoyle (1968) has found a general expansion for var $(\tilde{\theta})$ when $\tilde{\theta}$ is given by Neyman and Scott's series. The elimination of bias when retransforming estimates of this nature is an interesting problem area where further work is required.

EXERCISES

8.1 Obtain an approximation for the hypergeometric in terms of the binomial by equating the first two moments and adjusting the parameters to retain the index as a positive integer, while the means are held equal. (Sandiford, 1960)

192

8.2   Show, by an example, that the approximation suggested in Exercise 8.1 does not necessarily preserve symmetry.

(Ord, 1968a)

8.3   Derive an expansion of type (8.5) for the Poisson distribution with parameter λ in terms of the normal, with standard deviate $z = (r - \lambda)\lambda^{-\frac{1}{2}}$. Hence obtain an expansion for the random variable, such as (8.10). What modifications could be made to z to improve the accuracy of the approximation?

8.4   Show that the Wilson–Hilferty approximation for the gamma can be inverted to give the approximation, for gamma quantile x_P, as

$$x_P = n\{1 + z_P(2/9n)^{\frac{1}{2}} - (2/9n)\}^3,$$

where z_P is the corresponding normal quantile.

(Wilson & Hilferty, 1931)

8.5   Using relations (8.14) to (8.16) and the Peizer–Pratt approximations for the Poisson and the binomial, derive approximations for the gamma, beta and negative binomial distributions.

(Peizer & Pratt, 1968)

8.6   For the generalised Poisson models with pgf $\exp[\lambda\{g(z) - 1\}]$, where $g(z) = \Sigma\, p_i z_i$, the pgf of the generaliser, show that the limiting forms corresponding to (a) λ large, (b) λ small, and (c) $g(z) = p_0 + p_1 z + o(p_1)$ lead to the limits listed in section 8.6 for the Neyman Type A.

8.7   Show that, when variance-stabilising transforms are applied separately to two jointly distributed random variables, their correlation structure is unchanged, asymptotically.

APPENDIX A

ORTHOGONAL POLYNOMIALS

The purpose of this appendix is to outline the basic theory and properties of orthogonal polynomials in so far as these are used in the main
text. For a fuller discussion see Szegö (1939) and Erdelyi (1953). The
Stieltjes form of the integral is used, to save repeating each step for
continuous and discrete distributions; that is

$$\begin{aligned}
\int_T dW(x) &= \int_T w(x)dx, \quad x \text{ continuous} \\
&= \sum_T w(x), \qquad x \text{ discrete}
\end{aligned} \right\} \tag{A1}$$

where $w(x)$ is the weighting function, proportional to the density
function $f(x)$. All integrals are taken over T, the range of X. If $\int x^{2m}dW$
exists, we may define the polynomials $q_j = q_j(x)$, $j = 0, 1, \dots , m$,
such that q_j is of order j or less in x (not a necessary condition, but
one which suits our present purpose), and

$$\begin{aligned}
\int q_i q_j \, dW &= 0, \quad i \neq j, \\
&= c_j, \quad i = j,
\end{aligned} \right\} \tag{A2}$$

where the c_j are constants. If we take $q_0 = 1$, $dW \geqslant 0$ and $\int dW = 1$,
$W(x)$ is a cumulative distribution function, so that the $\{q_j\}$ correspond
to some distribution. Using (A2), we may define the polynomials successively, and since q_j is of the form

$$a_{j0} + a_{j1} x + \dots + a_{jj} x^j, \tag{A3}$$

$(j + 1)$ coefficients have to be fixed. Equation (A2) gives j conditions
on the $\{a_{ij}\}$, leaving c_j to be determined. Convenient choices are $c_j =
j!$ or $c_j = 1$.

When $w(x)$ is continuous, the classical orthogonal polynomials
(listed in Table A1) are characterised as follows:

if $\{q_j(x)\}$ is a system of orthogonal polynomials of form (A3),
obeying (A2), then

(a) $\{q_j'(x)\}$ is also a system of orthogonal polynomials;

(b) for all j, $q_j(x)$ satisfies

$$A(x)q'' + B(x)q' + \lambda_j q = 0, \quad j = 0, 1, \ldots ,$$

where A and B are independent of j, and λ_j is independent of x;

(c) the $\{q_j\}$ are generated by the Rodrigues formula

$$q_j(x) = \frac{1}{K_j w(x)} D^j \{w(x)\, P^j\}, \tag{A4}$$

where P is a polynomial in x, independent of j, the K_j are constants, and D is the operator d/dx.

To define the classical system, consider (A4) when $j = 1$, which becomes

$$\frac{w'(x)}{w(x)} = \frac{K_1 q_1 - P'}{P}. \tag{A5}$$

It is readily shown (Erdelyi, 1953) that if P is of higher order than 2 in x, (A2) and (A4) cannot hold simultaneously; when P is a quadratic, (A5) is clearly the Pearson differential equation [Chapter 1, equation (1.4)]. That is, the orthogonal polynomials generated by (A4) correspond to the Pearson curves; these are listed in Table A1.

Table A1

Classical continuous orthogonal polynomials

Name	Pearson type	Range	$w(x)$	P	Comments
Legendre (spherical)	Uniform (II)	$[-1, 1]$	1	$1 - x^2$	
Gegenbauer (ultra-spherical)	II	$[-1, 1]$	$(1 - x^2)^{m - \frac{1}{2}}$	$1 - x^2$	Chebyshev when $m = 0$ or 1.
Jacobi	I	$[-1, 1]$	$(1 - x)^a (1 + x)^b$	$1 - x^2$	II, VIII, IX and XII are special cases.
Hermite	Normal	$(-\infty, \infty)$	$\exp\left(-\tfrac{1}{2} x^2\right)$	1	
Laguerre	III, X	$[0, \infty)$	$x^a e^{-x}$	x	X when $a = 0$.
Hildebrandt	IV, VII	$(-\infty, \infty)$	$(1 + x^2)\, e^{\nu\theta}$ $\theta = \tan^{-1} x$	$1 + x^2$	Only first $(m - 1)$ polynomials exist.
	V	$[0, \infty)$	$x^{-2m} \exp(-1/x)$	x^2	
	VI, XI	$[0, \infty)$	$x^a (1 + x)^{-a-2m}$	$x(1 + x)$	

The most important set is the Hermite polynomials, $H_j(x)$, related to the normal. The first six are listed in section 2.1, while the general form is

$$H_j(x) = x^j - \frac{j^{(2)} x^{j-2}}{2 \cdot 1!} + \frac{j^{(4)} x^{j-4}}{2^2 \cdot 2!} - \frac{j^{(6)} x^{j-6}}{2^3 \cdot 3!} + \ldots, \qquad (A6)$$

P being taken as -1 rather than $+1$, by convention. It follows directly from (A4) that

$$H_j(x) \, w(x) = (-D)^j \, w(x), \qquad (A7)$$

while from the properties of the Laplace transform, we have the generating function for the H_j:

$$\exp{(tx - \tfrac{1}{2}t^2)} = \sum_{j=0}^{\infty} t^j H_j(x)/j! \,. \qquad (A8)$$

Differentiating (A8) and comparing coefficients,

$$D\{H_j(x)\} = jH_{j-1}(x) \qquad (A9)$$

and

$$D^k\{H_j(x)\} = j^{(k)}H_{j-k}(x), \quad j \geqslant k, \qquad (A10)$$

$$= 0, \text{ otherwise,}$$

demonstrating condition (a) above.

Differentiating (A7) and using (A9) yields

$$H_j(x) - xH_{j-1}(x) + (j-1)H_{j-2}(x) = 0; \qquad (A11)$$

further application of (A9) producing condition (b),

$$D^2 H_j(x) - xDH_j(x) + jH_j(x) = 0. \qquad (A12)$$

The classical discrete polynomials follow in the same way, on replacing D by the difference operator Δ. The Rodrigues formula becomes

$$q_j(x) = \frac{1}{K_j \, w(x)} \, \Delta^j \, \{w(x-j)P(x)\ldots P(x-j+1)\} \qquad (A13)$$

and the density functions for which (A13) holds satisfy the difference equation (5.1) in Chapter 5, the discrete analogue of (A5) above. These polynomials are listed in Table A2.

Table A2

Classical discrete orthogonal polynomials

Name	Discrete distribution	Range	$w(x)$	P
Chebyshev	Rectangular	$[0, n-1]$	1	$x(n-x)$
Krawtchouk	Binomial	$[0, n]$	$\binom{n}{x} p^x q^{n-x}$	$p(n-x+1)$
Charlier	Poisson	$[0, \infty)$	$a^x/x!$	1
Meixner	Pascal	$[0, \infty)$	$(b)_x c^x/x!$	$c(b+x-1)$
Hahn	Hypergeometric	$[0, n]$ or $[0, \infty)$	$\dfrac{(b)_x (c)_x}{(a)_x x!}$	$(b+x-1)(c+x-1)$

Note: $(b)_x = (b+x-1)\ldots(b+1)b$.

TABLES AND CHARTS

The purpose of this appendix is to provide a list of comprehensive, readily available tables and charts relevant to the various distributions discussed in the main text. No attempt has been made to list all the tables available on any particular topic, although some lists are fairly complete. In general, the primary source has not been quoted if the secondary source is widely available. Wherever possible, at least one set of general tables (listed in section 1, below), which contains the item of interest, has been given; usually the Pearson–Hartley tables, perhaps because of their historical attachment to the Pearson system of curves.

Notation

mD and mS denote m decimal places and m significant figures, respectively. P, or $P(x)$, denotes the distribution function, and f, or $f(x)$, the density function. For confidence intervals,

$$1 - 2\alpha = \text{prob } (x_L \leqslant x \leqslant x_U).$$

Other notation is introduced as required.

Most sections are subdivided into three parts for ease of reference:

(1) tables of P and f;
(2) tables of percentage points and
    confidence intervals;
(3) related tables of interest.

1 General works of reference and volumes of statistical tables

A code is given in round brackets after some works, for reference in later sections. Extracts from the more common tables appear in most, if not all, of the volumes of tables.

Abramowitz, M. and Stegun, I.A. (1964), *Handbook of Mathematical Functions*. New York: Dover. (AS)
Fisher, R.A. and Yates, F. (1963), *Statistical Tables for Biological, Agricultural and Medical Research*. Edinburgh: Oliver and Boyd. (FY)

Fletcher, A., Miller, J.C.P., Rosenhead L. and Comríe, L.J. (1962), *An Index of Mathematical Tables*. Oxford: Blackwell. 2 vol. Especially Chapters 14 and 15 of volume 1.

Greenwood, J.A. and Hartley, H.O. (1962), *Guide to Tables in Mathematical Statistics*. Oxford: Oxford University Press.

Hald, A. (1952), *Statistical Tables and Formulas*. New York: Wiley. (Hald).

Johnson, N.L. and Kotz, S. (1969–70) – see section 1 of main list of references. Contains lists of tables for each major distribution.

Lindley, D.V. and Miller, J.C.P. (1961), *Elementary Statistical Tables*. Cambridge: Cambridge University Press.   (LM)

Mayne, A.J. (1972), *Handbook of Statistical Distributions* (to appear). Contains extensive lists of charts and tables.

Owen, D.B. (1962), *Handbook of Statistical Tables*. Reading, Mass.: Addison-Wesley.   (Owen)

Patil, G.P. and Joshi, S.W. (1968), *Dictionary and Bibliography of Discrete Distributions*. Edinburgh: Oliver and Boyd. Chapter 3 includes lists of tables classified by type of distribution.

Pearson, E.S. and Hartley, H.O. (1966), *Biometrika Tables for Statisticians*, Cambridge: Cambridge University Press. Volume 1, 3rd edn.   (PH)

Pearson, K. (1931), *Biometrika Tables for Statisticians*. Cambridge: Cambridge University Press. 2 volumes.   (KP1, KP2)

Rao, C.R., Mitra S.K., and Matthai, A. (1966), *Formulae and Tables for Statistical Work*. Calcutta: Statistical Publishing Society.

Rohlf, F.J. and Sokal, R.R. (1969), *Statistical Tables*. San Francisco: W.H. Freeman.

2 Pearson curves

Many of the tables relating to Pearson curves which appeared in KP1 and KP2 are no longer used, and only those of current interest are listed below.

2.1 For tables of P and f for various distributions, see sections 3.1 to 6.1, below.

2.2 Johnson, N.L. *et al.* (1963), Tables of percentage points of Pearson curves, for given $\sqrt{\beta_1}$ and β_2, expressed in standard measure, *Biometrika*, **50**, 459–98.
   For P, $1 - P = 0\cdot5$, $0\cdot25$, $0\cdot1$, $0\cdot05$, $0\cdot025$, $0\cdot01$, $0\cdot005$, $0\cdot0025$, 0 (if finite end-point);
   $\sqrt{\beta_1} = 0(0\cdot1)2$, β_2 in range $1\cdot6(0\cdot2)14\cdot4$,   4S.

These tables supersede those published earlier; see PH, Table 42.

2.3 KP1, Table LIV: Tables of functions used in determining the ordinate of a Pearson Type IV curve.

Simpson, J.A. and Welch, B.L. (1960), Tables of the bounds of the probability integral when the first four moments are given, *Biometrika*, **47**, 399–410.

Upper and lower bounds for $P(X \leqslant a)$, X in standard measure: for $\sqrt{\beta_1} = 0\,(1)\,4,\ 8;\ \Delta = \beta - \beta - 1 = 0,\ 0{\cdot}5,\ 1\,(1)\,4\,(2)$
$\qquad\qquad 8\,(4)\,16,\ \infty;$
$-a,\ a = 0\,(0{\cdot}5)\,3\,(1)\,10,\quad 3\mathrm{D}.$

3 The normal curve

Tables are universally available and a general guide to earlier sets is:

National Bureau of Standards (1951), *A Guide to Tables of the Normal Probability Integral*, Applied Maths. Series No. 21. Washington: U.S. Govt. Printing Office.

3.1 A few examples are:

AS, Table 26.1: P and f for $x = 0\,(0{\cdot}02)\,3\,(0{\cdot}05)\,5,\quad 15\mathrm{D}.$

Harvard University (1952), *Tables of the Error Function and its First Twenty Derivatives*. Cambridge, Mass.: Harvard University Press.   (HU 52)

$P - \frac{1}{2}$, f for $x = 0\,(0{\cdot}004)\,6{\cdot}468,\quad 6\mathrm{D}.$

PH, Table 1: P and f for $x = 0\,(0{\cdot}01)\,4{\cdot}5$, 7D and 1st and 2nd differences; for $x = 4{\cdot}5\,(0{\cdot}01)\,6{\cdot}0,\quad 10\mathrm{D}.$

Table 2: $-\ln(1-P)$ for $x = 5\,(1)\,50\,(10)\,100\,(50)\,500,\quad 5\mathrm{D}.$

Sheppard, W.F. (1939), *The Probability Integral*, British Association for the Advancement of Science, Math. Tables VII. Cambridge: Cambridge University Press.

$(2P-1)/f$ for $x = 0\,(0{\cdot}01)\,10,\quad 12\mathrm{D}.$
$\qquad\qquad\quad x = 0\,(0{\cdot}1)\,10,\quad 24\mathrm{D}.$

3.2 Kelley, T.L. (1948), *The Kelley Statistical Tables*. Cambridge, Mass.: Harvard University Press.

x for $P = 0{\cdot}5\,(0{\cdot}0001)\,0{\cdot}9999$, with corresponding values for f,   8D.

3.3 AS, Table 26.1: $f^{(j)}(x)$ for $j = 1$ (15D), 2 (10D), 3 (1) 6,   (8D), $x = 0\,(0{\cdot}02)\,3\,(0{\cdot}05)\,5.$

HU52 (see section 3.1, above): $f^{(j)}(x)$ for $j = 1$ (1) 10,
$x = 0$ (0·004) 8·236,   10D;
for $j = 11$ (1) 20, $x = 0$ (0·002) 10·902,   7S.

KP1, Table XXIX: $\tau_j = (j!)^{-\frac{1}{2}} (-1)^j f^{(j)}(x)$ for $j = 1$ (1) 6,
x for $1 - P = 0·001$ (0·001) 0·5,   5D.

KP2, Table VII: for $j = 1$ (1) 19, $x = 0$ (0·1) 4,   7D.

Tables of percentage points for a variety of derived statistics
(e.g. range, largest individual) are also included in most sets of tables.

4   The gamma distribution (see also section 9, *below*, on the Poisson).

$P \equiv P(x, j) = \int_0^x u^{j-1}e^{-u}du/\Gamma(j)$.   For $x = \frac{1}{2}\chi^2$, $P(x, j)$ represents
the chi-square distribution with $n = 2j$ degrees of freedom.

4.1   (a) *Tables*

Harter, H.L. (1964), *New Tables of the Incomplete Gamma
Function and of the Percentage Points of the Chi-
Squared and Beta Distributions*. Washington: Office
of Aerospace Research, USAF.
P for $y = x(j + 1)^{-\frac{1}{2}}$; $y = 0$ (0·1) 19·9, $j - 1 = 0·5$ (0·5)
73 (1) 163,   9D.

Khamis, S.H. and Rudert, W. (1965), *Tables of the Incomplete
Gamma Function Ratio*. Darmstadt: Von Liebig.
P for $y = 2x$; $y = 0$ (0·0001) 0·001 (0·001) 0·01 (0·01) 1
(0·05) 6 (0·1) 16 (0·5) 66 (1) 166 (2) 250; $j = 0·05$ (0·05)
10 (0·1) 20 (0·25) 70,   10D.

PH, Table 7: $1 - P$ for $y = 2x$; $y = 0·001$ (0·001) 0·01 (0·01)
0·1 (0·1) 6 (0·2) 10 (0·5) 20 (1) 120; $2j$ in range 1 (1) 70,
5D.

Pearson, K. (1934), *Tables of the Incomplete Gamma Function*.
Cambridge: Cambridge University Press.
$1 - P$ for $y = x(j + 1)^{-\frac{1}{2}}$; $y = 0$ (0·1) until $1 - P < 5 \times 10^{-8}$,
$j - 1 = -1$ (0·05) 0 (0·1) 5 (0·2) 50, 7D; $y = 0$ (0·1) 6,
$j - 1 = -1$ (0·01) − 0·75, 5D.

(b) *Charts*

Boyd, W.C. (1965), A nomogram for chi-square, *J. Amer.
Statist. Ass.*, **60**, 344−6.
P for $0·1 \leqslant x \leqslant 120$, $1 \leqslant 2j \leqslant 120$, $10^{-7} \leqslant P \leqslant 0·2$.

Crow, J.F. (1945), A chart for the χ^2 and t distributions, *J. Amer. Statist. Ass.*, **40**, 376.

Smirnov, N.V. and Potapov, M.K. (1957), A nomogram for the incomplete Γ-function and the χ^2 probability distribution, *Theor. Prob. Appl.*, **2**, 461–5.
P for $|y| \leqslant 3$, $y = (2x)^{\frac{1}{2}} - (2j)^{\frac{1}{2}}$, $1 \leqslant x \leqslant 30$, $0 \cdot 001 \leqslant P \leqslant 0 \cdot 999$, 2 or 3D.

4.2 Tables of percentage points are widely available. The most comprehensive set is probably –

Harter (1964). see section 4.1, *above*, for $n = 2j = 1\,(1)\,150$ $(2)\,330$;
$P, 1 - P = 0 \cdot 0001,\ 0 \cdot 0005,\ 0 \cdot 001,\ 0 \cdot 005,\ 0 \cdot 01,\ 0 \cdot 025,$ $0 \cdot 05,\ 0 \cdot 1\,(0 \cdot 1)\,0 \cdot 5,$ 6D.

4.3 Tables of the digamma function are needed to evaluate the ML estimator for j.

British Association for the Advancement of Science (1951), *The Gamma and Polygamma Functions*. Cambridge: Cambridge University Press.

Davies, H.T. (1933, 1935), *Tables of the Higher Mathematical Functions*. Bloomington, Indiana: Principia Press. 2 vol.

Pairman, E. (1916), *Tables of the Digamma and Trigamma Functions*. Biometrika Tracts for Computers, No. 1. Cambridge: Cambridge University Press.

5 The beta and F distributions (see also section 10, *below*, on the binomial).

$$P \equiv P(x, a, b) = \int_0^x u^{a-1}(1-u)^{b-1}\,du/\beta(a, b).$$ For F with degrees of freedom $n_1 = 2a$, $n_2 = 2b$ use $y = x/(1 - x)$. x).

5.1 Pearson, K. (1934), *Tables of the Incomplete Beta Function*. Cambridge: University Press (new edition by Pearson, E.S. and Johnson, N.L., 1968).
P for $x = 0 \cdot 01\,(0 \cdot 01)\,1$; $a, b = 0 \cdot 5\,(0 \cdot 5)\,11\,(1)\,50$, $a \geqslant b$, 7D.

PH, Table 17: A chart for the incomplete beta function and the cumulative binomial distribution.
Graduations at $a = 2\,(1)\,10\,(2 \cdot 5)\,25\,(5)\,50\,(10)\,100\,(25)\,150,$ 200;

202

$b = 1\ (1)\ 4,\ 0\!\cdot\!05 \leqslant x \leqslant 0\!\cdot\!999\ 95;$
$b = 4\ (1)\ 15,\ 0\!\cdot\!10 \leqslant x \leqslant 0\!\cdot\!990;$
$b = 15,\ 20,\ 30,\ 60,\ 0\!\cdot\!30 \leqslant x \leqslant 0\!\cdot\!94,\quad 2D.$
P running from $0\!\cdot\!0001$ to $0\!\cdot\!9999.$

5.2 Harter, H.L. (1964), see section 4.1, *above*.
$a,\ b = 1\ (1)\ 40;$
$P,\ 1 - P = 0\!\cdot\!0001,\ 0\!\cdot\!0005,\ 0\!\cdot\!001,\ 0\!\cdot\!025,\ 0\!\cdot\!05,\ 0\!\cdot\!1\ (0\!\cdot\!1)\ 0\!\cdot\!5,$
7D.

Thompson, C.M. (1941), Tables of percentage points of the incomplete beta function, *Biometrika*, **32**, 151–81.
$2b = 1\ (1)\ 30,\ 40,\ 60,\ 120,\infty;$
$2b = 1\ (1)\ 10,\ 12,\ 15,\ 24,\ 30,\ 40,\ 60,\ 120,\infty;$
$P = 0\!\cdot\!005,\ 0\!\cdot\!01,\ 0\!\cdot\!025,\ 0\!\cdot\!1,\ 0\!\cdot\!25,\ 0\!\cdot\!5,\quad 5D.$

Reproduced as PH, Table 16 and elsewhere, and inverted to give per cent points for F, as in PH, Table 18.

6 The Student t distribution

$$P(t, n) \propto \int_{-\infty}^{t} (1 + x^2/n)^{-\frac{1}{2}(n+1)}\,dx.$$

6.1 PH, Table 9: for $n = 1\ (1)\ 20;\ t = 0\ (0\!\cdot\!1)\ 4\ (0\!\cdot\!2)\ 8;$
$n = 20\ (1)\ 24,\ 30,\ 40,\ 60,\ 120;\ t = 0\ (0\!\cdot\!05)\ 2\ (0\!\cdot\!1)\ 4,\ 5,\quad 5D.$

Boyd, W.C. (1969), A nomogram for the Student–Fisher t test, *J. Amer. Statist. Ass.*, **64**, 1664–7.
Nomogram for $n = 1\ (1)\ 5,\ 10,\ 20,\infty;\ P = 0\!\cdot\!000\ 01$ to $0\!\cdot\!05;$
$t = 1$ to 60, 3S.

Crow, J.F. (1945), A chart for the χ^2 and t distributions, *J. Amer. Statist. Ass.*, **40**, 376.

James-Levy, P.E. (1956), A nomogram for the integral law of Student's t distribution, *Theor. Prob. Appl.*, **1**, 246–8.
Graduations at $n = 3\ (1)\ 20\ (5)\ 50\ (10)\ 100\ (20)\ 200\ (100)\ 500,$
1000;
$1\!\cdot\!3 \leqslant t \leqslant 13,\ 0\!\cdot\!9 \leqslant P \leqslant 0\!\cdot\!9995,\quad 2$ or 3D.

6.2 PH, Table 12 (for example):
$n = 1\ (1)\ 30,\ 40,\ 60,\ 120,\infty;$
$1 - P = 0\!\cdot\!4,\ 0\!\cdot\!25,\ 0\!\cdot\!1,\ 0\!\cdot\!05,\ 0\!\cdot\!025,\ 0\!\cdot\!01,\ 0\!\cdot\!005,\ 0\!\cdot\!0025,$
$0\!\cdot\!001,\ 0\!\cdot\!0005,\quad 3S.$
Extended by –

Federighi, E.T. (1959), Extended tables of the percentage points of the Student's t distribution, *J. Amer. Statist. Ass.*, **54**, 683–8.

$$\left. \begin{array}{l} \text{for } 1 - P = 0{\cdot}25 \times 10^{-m} \\ \phantom{\text{for } 1 - P} = 0{\cdot}1 \ \times 10^{-m} \end{array} \right\} m = 0 \ (1) \ 6$$

$$= 0{\cdot}05 \times 10^{-m}, \ m = 0 \ (1) \ 5$$

$n = 1 \ (1) \ 30 \ (5) \ 60 \ (10) \ 100, \ 200, \ 500, \ 1000, \ 2000, \ 10\ 000, \quad 3D.$

7 The Johnson system

7.1 For the Johnson curves, tables of transforms given in section 7.3 *below* are used in conjunction with normal tables.

7.2 Johnson, N.L. (1964), Tables of percentage points of S_U curves, given $\sqrt{\beta_1}$ and β_2, expressed in standard measure, *Univ. N. Carolina Mimeo Series* No. 408.

For P, $1 - P = 0{\cdot}0005, \ 0{\cdot}001, \ 0{\cdot}0025, \ 0{\cdot}005, \ 0{\cdot}01, \ 0{\cdot}025,$ $0{\cdot}05, \ 0{\cdot}10, \ 0{\cdot}25, \ 0{\cdot}5.$

For $\sqrt{\beta_1}$ at intervals of $0{\cdot}1$, β_2 at intervals of $0{\cdot}2$.

7.3 Johnson, N.L. (1949), Systems of frequency curves generated by methods of translation, *Biometrika*, **36**, 149–76.

Abac to facilitate fitting of S_U curves; now superseded by –

Johnson, N.L. (1965), Tables to facilitate fitting S_U frequency curves, *Biometrika*, **52**, 547–58.

Gives values of γ and δ (defined in Chapter 2, *above*, section 2.12) for

$\sqrt{\beta_1} = 0{\cdot}05 \ (0{\cdot}05) \ 2$, β_2 in range $3{\cdot}2 \ (0{\cdot}1) \ 8 \ (0{\cdot}2) \ 15{\cdot}0$.

Yuan, P.T. (1933), On the logarithmic frequency distribution, *Ann. Math. Statist.*, **4**, 30–74.

Tables to aid fitting of lognormal, using the first three moments (for discussion, see Chapter 2, *above*, Section 2.13).

Transformations

Harvard Computational Laboratory (1949), *Tables of Inverse Hyperbolic Functions*. Cambridge, Mass.: Harvard University Press.

arctanh x, $0 \leqslant x \leqslant 1$;

arcsinh x, $0 \leqslant x < 3{\cdot}5$, $3{\cdot}5 \leqslant x \leqslant 22\ 980$, various intervals, 9D.

Extracts from these tables appear as AS, Table 4.17.

LM, Table 4: arctanh x ·for x = 0 (0·02) 0·8 (0·01) 0·94 (0·001)
0·999,   3D.

PH, Table 14: arctanh x for x = 0 (0·002) 0·998, 4D, with proportional parts for fourth decimal place, $0·25 \leqslant x \leqslant 0·75$.

8 The bivariate normal distribution

$\Pr(X \geqslant h, Y \geqslant k; \rho) = L(h, k, \rho); \Pr(X \geqslant h) = Q(h).$

$V(h, ah) = \frac{1}{4} + L(h, 0, \rho) - L(0, 0, \rho) - \frac{1}{2}Q(h),$
   where $\rho = -a(1 + a^2)^{-\frac{1}{2}}$.

$L(h, k, \rho) = L\{h, 0, w(\rho h - k)s(h)\} + L\{k, 0, w(\rho k - h)s(k)\} - \delta,$
where $w^{-2} = h^2 + k^2 - 2\rho hk,$ $s(h) \equiv$ sign of h,

   δ = 0, if $hk > 0$, or $hk = 0, h + k \geqslant 0$,

   $= \frac{1}{2}$, otherwise.

National Bureau of Standards (1959), *Tables of the Bivariate*
Normal Distribution Function and Related Tables.
Applied Maths. Series No. 50. Washington: U.S. Govt.
Printing Office.
  $1 - P(h, k, \rho)$ for h, k = 0 (0·1) 4, ρ = 0 (0·05) 0·95 (0·01) 1,
6D.

   h, k = 0 (0·1) A, $-\rho$ = 0 (0·05) 0·95 (0·01) 1,
   where A is such that $P < 5 \times 10^{-7}$,   7D.
  $V(h, ah)$ for a = 0·1 (0·1) 1, h = 0 (0·01) 4 (0·02) 5, 6,   7D.

Owen, D.B. (1957), *The Bivariate Normal Probability Function.*
Washington: Office of Technical Services, U.S. Dept. of
Commerce.
  $T(h, a) = (2\pi)^{-1}$ arctan $a - V(a, ah)$ for a = 0 (0·025) 1;
  h = 0 (0·01) 3·5 (0·05) 4·75,   6D.

KP2, Tables VIII and IX:
  $1 - P(h, k, \rho)$ for h, k = 0 (0·1) 2·6, ρ = 0 (0·05) 1,   6D.
                          $-\rho$ = 0 (0·05) 1,   7D.

Zelen, M. and Severo, N.C. (1960), Graphs for bivariate normal
probabilities, *Ann. Math. Statist.*, **31**, 619–24; reproduced
in AS, pp. 937–9.
  $L(h, 0, \rho)$ for $0 \leqslant h \leqslant 1, -1 \leqslant \rho \leqslant 0, 0·01 \leqslant P \leqslant 0·24$
                  $0 \leqslant \rho \leqslant 1, 0·07 \leqslant P \leqslant 0·48$
         $h \geqslant 1$ $-1 \leqslant \rho \leqslant 1, 0·01 \leqslant P \leqslant 0·15$,   3D.

9   The Poisson distribution (see section 4, *above*, on the gamma), parameter m.

9.1   G.E.C. Defence Systems Department (1962), *Tables of the Individual and Cumulative Terms of the Poisson Distribution.* New York: Van Nostrand.

P and f for $m = 10^{-7}$ (10^{-7}) $1 \cdot 5 \times 10^{-6}$ (5×10^{-7}) $1 \cdot 5 \times 10^{-5}$ (10^{-6}) 5×10^{-5} (5×10^{-6}) $0 \cdot 0005$ (10^{-5}) $0 \cdot 001$ (5×10^{-5}) $0 \cdot 005$ $(0 \cdot 0001)$ $0 \cdot 01$ $(0 \cdot 0005)$ $0 \cdot 2$ $(0 \cdot 001)$ $0 \cdot 4$ $(0 \cdot 005)$ $0 \cdot 5$ $(0 \cdot 01)$ 1 $(0 \cdot 05)$ 2 $(0 \cdot 1)$ 5 $(0 \cdot 5)$ 10 (1) 100 (5) 205,   8D.

Kitagawa, T. (1951), *Tables of the Poisson Distribution.* Tokyo: Baifukan.

f for $m = 0 \cdot 001$ $(0 \cdot 001)$ 1 $(0 \cdot 01)$ 5,   8D.
            5 $(0 \cdot 01)$ 10,   7D.

Molina, E.C. (1942), *Poisson's Exponential Binomial Limit.* New York: Van Nostrand.

P and f for $m = 0 \cdot 1$ $(0 \cdot 1)$ 16 (1) 100,   6D.
            $m = 0 \cdot 001$ $(0 \cdot 001)$ $0 \cdot 01$ $(0 \cdot 01)$ 3,   7D.

PH, Table 39: f for $m = 0 \cdot 1$ $(0 \cdot 1)$ 15,   6D.

9.2   Blyth, C.R. and Hutchinson, D.W. (1961), Table of Neyman – shortest unbiased confidence intervals for the Poisson parameter, *Biometrika*, **48**, 191–4.

$z = x + y$ for x Poisson, y uniform on $[0, 1]$, $\alpha = 0 \cdot 025$, $0 \cdot 005$;
$z = 0 \cdot 01$ $(0 \cdot 01)$ $0 \cdot 1$ $(0 \cdot 02)$ $0 \cdot 2$ $(0 \cdot 05)$ 1 $(0 \cdot 1)$ 10 $(0 \cdot 2)$ 40 $(0 \cdot 5)$ 55 (1) 250,   3S.

Crow, E.L. and Gardner, R.S. (1959), Confidence intervals for the expectation of a Poisson variable, *Biometrika*, **46**, 441–53.

$\alpha = 0 \cdot 1$, $0 \cdot 05$, $0 \cdot 025$, $0 \cdot 005$, $0 \cdot 0005$; $x = 1$ (1) 300,   5S.

PH, Table 40; $\alpha = 0 \cdot 05$, $0 \cdot 025$, $0 \cdot 01$, $0 \cdot 005$, $0 \cdot 001$; $x = 1$ (1) 50, 4S.

Walton, G.S. (1970), see section 10.2, *below*. Reports availability of new tables for $\alpha = 0 \cdot 25$ $(0 \cdot 05)$ $0 \cdot 05$, $0 \cdot 025$, $0 \cdot 005$; $x = 1$ (1) 100.

9.3   Barton, D.E., David, F.N. and Merrington, M. (1960), Tables for the solution of the exponential equation $\exp(-a) + ka = 1$, *Biometrika*, **47**, 439–45.

To find ML estimator for m when distribution is truncated at zero, solutions to $e^{-a} + ka = 1$ for $k = 0 \cdot 05$ $(0 \cdot 001)$ 1,   7D.

Grab, E.L. and Savage, I.R. (1954), Tables of the expected
value of $1/x$ for positive Bernoulli and Poisson variables,
J. Amer. Statist. Ass., **49**, 169–77.
$m = 0\cdot01,\ 0\cdot05\ (0\cdot05)\ 1\ (0\cdot1)\ 2\ (0\cdot2)\ 5\ (0\cdot5)\ 7\ (1)\ 10\ (2)\ 20,\quad 5D.$

For tables to assist in calculations for the hyper-Poisson,
see section 12.3, *below*.

10 The binomial distribution (see also section 5, *above*, on the beta).

Parameters n and p. The random variable is given over 0 (1) n unless
otherwise stated.

10.1 Harvard Computational Laboratory (1955), *Tables of the Cumulative Binomial Probability Distribution*. Cambridge, Mass.: Harvard University Press.
P for $n = 2\ (1)\ 50\ (2)\ 100\ (10)\ 200\ (20)\ 500\ (50)\ 1000;$
$p = 0\cdot01\ (0\cdot01)\ 0\cdot5,\quad 5D.$

National Bureau of Standards (1950), *Tables of the Binomial Probability Distribution*, Applied Maths. Series, No. 6. Washington, DC: U.S. Govt. Printing Office.
P and f for $n = 2\ (1)\ 49;\ p = 0\cdot01\ (0\cdot01)\ 0\cdot5,\quad 7D.$

PH, Table 37: f for $n = 5\ (5)\ 50;\ p = 0\cdot01,\ 0\cdot02\ (0\cdot02)\ 0\cdot1\ (0\cdot1)\ 0\cdot5,\quad 5D.$

Robertson, W.H. (1960), *Tables of the Binomial Distribution Function for Small Values of p*. Washington, DC: Office of Technical Services, U.S. Dept. of Commerce.
P for $n = 2\ (1)\ 100\ (2)\ 200\ (10)\ 500\ (20)\ 1000;$
$p = 0\cdot001\ (0\cdot001)\ 0\cdot02,$
and $n = 2\ (1)\ 50\ (2)\ 200\ (10)\ 500\ (20)\ 1000;$
$p = 0\cdot021\ (0\cdot001)\ 0\cdot05,\quad 5D.$

Romig, H.G. (1953), *50–100 Binomial Tables*. New York: Wiley.
P and f for $n = 50\ (5)\ 100;\ p = 0\cdot01\ (0\cdot01)\ 0\cdot5,\quad 6D.$

U.S. Ordnance Corps (1952), *Tables of Cumulative Binomial Probabilities*, ORDP 20–1. Washington, DC: Office of Technical Services.
P for $n = 1\ (1)\ 150;\ p = 0\cdot01\ (0\cdot01)\ 0\cdot5,\quad 7D.$

Weintraub, S. (1963), *The Cumulative Binomial Probability Distribution for Small Values of p*. Illinois: Free Press of Glencoe. London: Collier-Macmillan.

P for $n = 1$ (1) 100; $p = 0\cdot000\,01,\ 0\cdot0001\ (0\cdot0001)\ 0\cdot001$
$(0\cdot001)\ 0\cdot1,\quad 10D.$

10.2 Blyth, C.R. and Hutchinson, D.W. (1960), Table of Neyman—
shortest unbiased confidence intervals for the binomial
parameter, *Biometrika*, **47**, 381–91.
$\alpha = 0\cdot025,\ 0\cdot005;\ n = 2$ (1) 24 (2) 50, $\quad 0 \leqslant r + y \leqslant m,$
$m = \left[\dfrac{n+1}{2}\right];\ y$ uniform on $[0,\ 1],\quad 2S.$

Crow, E.L. (1956), Confidence limits for a proportion, *Biometrika*, **43**, 423–35.
$\alpha = 0\cdot05,\ 0\cdot025,\ 0\cdot005;\ n = 1$ (1) 30, $\quad r = 0$ (1) $n,\quad 3D.$

Pachares, J. (1960), Tables of confidence intervals for the binomial distribution, *J. Amer. Statist. Ass.*, **55**, 521–33.
$\alpha = 0\cdot05,\ 0\cdot025,\ 0\cdot01.\ 0\cdot005;\ n = 55$ (5) 1000; $r = 0$ (1)
$n - 1,\quad 4D.$

PH, Table 41, Chart providing confidence intervals for p in
binomial sampling.
$\alpha = 0\cdot025,\ 0\cdot005;\ r/n = 0$ (0$\cdot$02) $0\cdot5$, plotted for $n = 8$ (2) 12
(4) 24, 30, 40, 60, 100, 200, 400, 1000, $\quad 2$ or 3D.

Walton, G.S. (1970), A note on nonrandomised Neyman—shortest
unbiased confidence intervals for the binomial and
Poisson parameters, *Biometrika*, **57**, 223–4.
Reports availability of new tables for $\alpha = 0\cdot25$ (0$\cdot$05) $0\cdot05$,
$0\cdot025,\ 0\cdot005;\ n = 1$ (1) 30.

10.3 Grab, E.L. and Savage, I.R. (1954), Tables of the expected
value of $1/x$ for positive Bernoulli and Poisson variables,
J. Amer. Statist. Ass., **49**, 169–177.
$n = 2$ (1) 30; $p = 0\cdot01,\ 0\cdot05$ (0$\cdot$05) $0\cdot95,\ 0\cdot99,\quad 5D.$

Miller, J.C.P. (ed.) (1954), *Tables of Binomial Coefficients*,
Math. Tables, Vol. 3. Cambridge: Cambridge University Press.
$\dbinom{n}{r}$ for $r \leqslant \tfrac{1}{2}n \leqslant 100$ plus supplementary tables up to
$n = 5000$ for small r.

11 The negative binomial and logarithmic series distributions
(parameters q and k)

11.1   Brown, B. (1965), *Some Tables of the Negative Binomial Distribution and their Use.* Santa Monica, California: Rand Corporation.

P and f for mean, $\mu = 0 \cdot 25 \ (0 \cdot 25) \ 1 \ (1) \ 10$

$$(1 - q)^{-1} = 1 \cdot 5 \ (0 \cdot 5) \ 5 \ (1) \ 7, \quad 4D.$$

Patil, G.P., Kamat, A.R. and Wani, J.K. (1964), *Certain Studies on the Structure of the Logarithmic Series Distribution and Related Tables.* Ohio: Aerospace Research Laboratories.

Williamson, E. and Bretherton, M.H. (1963), *Tables of the Negative Binomial Probability Distribution.* New York: Wiley.
$q = 0 \cdot 05, \ 0 \cdot 1 \ (0 \cdot 02) \ 0 \cdot 90, \ 0 \cdot 95;$
$k = 0 \cdot 1 \ (0 \cdot 1) \ 2 \cdot 5 \ (0 \cdot 5) \ 10$, various up to 200,   6D.

Williamson, E. and Bretherton, M.H. (1964), Tables of the logarithmic series distribution, *Ann. Math. Statist.*, **35**, 284–97.
P and f for mean, $\mu = 1 \cdot 1 \ (0 \cdot 1) \ 2 \ (0 \cdot 5) \ 5 \ (1) \ 10$ for $P \leqslant 0 \cdot 999$, 5D.
ML est. $\hat{q}$ for $\mu = 1 \ (0 \cdot 1) \ 10 \ (1) \ 50$,   5D.

11.2   Patil, G.P. and Wani, J.K. (1965), Maximum likelihood estimation for the complete and truncated logarithmic series distributions, in G.P. Patil (ed.), *Classical and Contagious Discrete Distributions.* Calcutta: Statistical Publishing Company, 398–409.
ML est. $\hat{q}$ for $\mu = 1 \cdot 02 \ (0 \cdot 02) \ 2 \ (0 \cdot 05) \ 4 \ (0 \cdot 1) \ 8 \ (0 \cdot 2) \ 16 \ (0 \cdot 5)$ 30 (2) 40 (5) 60 (10) 140 (20) 200.
For complete range and for truncation at upper end-point,
$d = 4 \ (1) \ 8 \ (2) \ 12 \ (5) \ 40 \ (10) \ 60 \ (20) \ 100, \ 200, \ 500, \ 1000, \quad 4D.$

12 Hypergeometric distributions

12.1   Lieberman, G.J. and Owen, D.B. (1961), *Tables of the Hypergeometric Probability Distribution.* Stanford: Stanford University Press.
P and f for $N = 2 \ (1) \ 50 \ (10) \ 100$, $M = 1 \ (1) \ N - 1$, $n = 1 \ (1) \ M$;
$r = 0 \ (1) \ n, \ N \leqslant 25, \ r = 0 \ (1) \left[\dfrac{n+1}{2} \right], \ N > 25, \quad 6D;$
plus more restricted tables for large N.

12.2 Chung, J.H. and DeLury, D.B. (1950), *Confidence Limits for the Hypergeometric Distribution*. Oxford: Oxford & Toronto University Presses.

α = 0·05, 0·025, 0·005, charts for r/n and N = 500, 2500, 10 000;

b = n/N = 0·05 (0·01) 0·1 (0·1) 0·9, 2D.

12.3 Rushton, S. and Lang, E.D. (1954), Tables of the confluent hypergeometric function, *Sankhyā*, **15**, 377−411.

$M(a, b, x)$, a = 0·5 (0·5) 40; b = 0·5 (0·5) 3·5; x = 0·02 (0·02) 0·1 (0·1) 1 (1) 10 (10) 50, 100, 200, 7S.

Slater, L.J. (1960), *Confluent Hypergeometric Functions*. Cambridge: Cambridge University Press.

$M(a, b, x)$, a = −1 (0·1) 1, b = −1 (0·1) 1, x = −1 (0·1) 10, 8S.

$M(a, b, 1)$, a = −11 (0·2) 2, b = −4 (0·2) 1, 8S.

13 Other discrete distributions

Haight, F.A. and Breuer, M.A. (1960), The Borel−Tanner distribution, *Biometrika*, **47**, 143−50.

See section 5.12 in main text, equation (5.44).

P for α = 0·01 (0·01) 0·62, while $P \leqslant$ 0·999, 5D.

Haight, F.A. (1961), A distribution analogous to the Borel−Tanner, *Biometrika*, **48**, 167−73.

See section 5.12 in main text, equation (5.45).

P for α = 0·01 (0·01) 0·62, while $P \leqslant$ 0·999, 5D.

Sherbrooke, C.C. (1968), Discrete compound Poisson processes and tables of the geometric Poisson distribution, *Naval Res. Logist. Quart.*, **15**, 189−203 (Pólya−Aeppli).

Tables of f for $m = \lambda\tau(1 - \tau)^{-1}$ = 0·1, 0·25 (0·25) 1 (0·5) 3 (1) 10

$q = (1 + \tau)(1 - \tau)^{-1}$ = 1·5 (0·5) 5 (1) 7, $r \leqslant$ 30, 4D.

14 Variance stabilising transforms

14.1 $y(x) = \arcsin \sqrt{x}$.
FY: $y(x)$ for $x = 0$ $(0 \cdot 01)$ $0 \cdot 99$,   2D (degrees).

Hald: $2y(x)$ for $x = 0$ $(0 \cdot 001)$ $0 \cdot 999$,   4D (radians).

LM: $y(x)$ for $x = 0$ $(0 \cdot 01)$ $0 \cdot 99$,   3D (radians).

Mosteller, F. and Youtz, C. (1961), Tables of the Freeman–
   Tukey transformations for the binomial and Poisson
   distributions, *Biometrika*, **48**, 433–40.
   $\frac{1}{2}y\{r/(n + 1)\} + \frac{1}{2}y\{(r + 1)/(n + 1)\}$ for $n = 1$ (1) 150,
   $r = 0$ (1) n,   2D (degrees).

Owen: $2y(x)$ for $x = 0$ $(0 \cdot 0001)$ $0 \cdot 0045$ $(0 \cdot 001)$ $0 \cdot 995$, $0 \cdot 9955$
                  $(0 \cdot 0001)$ $0 \cdot 9998$,   4D (radians).

Stevens, W.L. (1953), Tables of the angular transformation,
   Biometrika, **40**, 70–73.
   $50 - 10\sqrt{10}\,y\{(n - 2r)/n\}$ for $r/n = 0$ $(0 \cdot 001)$ $0 \cdot 5$,   5D.
                                  $= 0$ $(0 \cdot 0001)$ $0 \cdot 02$,   5D.
   r/n as proper fraction, denominator $\leqslant 30$,   5D (all radians).

14.2 $y(x) = \operatorname{arcsinh} \sqrt{x}$.

Anscombe, F.J. (1950), Tables of the hyperbolic transform
   $\sinh^{-1} \sqrt{x}$, *J. Roy. Statist. Soc.*, *Ser. A*, **113**, 228–9.
   $y(x)$ for $x = 0$ $(0 \cdot 01)$ 1 $(0 \cdot 1)$ 10 (1) 200 (10) 590,   3D.

Beall, G. (1942), The transformation of data from entomological field experiments so that the analysis of variance
   becomes applicable, *Biometrika*, **32**, 243–62.
   $k^{-\frac{1}{2}} y(kr)$ for $k = 0$ $(0 \cdot 02)$ $0 \cdot 1$ $(0 \cdot 05)$ $0 \cdot 3$ $(0 \cdot 1)$ $0 \cdot 6$ $(0 \cdot 2)$ 1;
                  $r = 0$ (1) 50 (5) 100 (10) 300,   2D.

LM: $y(x)$ for $x = 0$ $(0 \cdot 01)$ 1 $(0 \cdot 1)$ 10 (1) 100,   3D.

REFERENCES

I RECENT MAJOR WORKS ON DISTRIBUTIONS

Elderton, Sir W.P. and Johnson, N.L. (1969), *Systems of Frequency Curves*. London: Cambridge University Press. (1.3, 1.7, 1.11)

> An updating of Elderton's *Frequency Curves and Correlation* covering the main systems of curves introduced in Chapters 1—3.

Johnson, N.L. and Kotz, S. (1969), *Discrete Distributions*. New York: Houghton Mifflin.

Johnson, N.L. and Kotz, S. (1970), *Continuous Distributions* (2 vol.). New York: Houghton Mifflin.

> This three-volume work summarises the properties of all the major distributions.

Mardia, K.V. (1970), *Families of Bivariate Distributions*. London: Griffin. (3.1, 3.8, 3.10, 7.1)

> Deals extensively with the continuous bivariate forms discussed in Chapter 3.

Mayne, A.J. (1972), *Approximations to Statistical Distributions*. (To appear.) (5.13, 8.2)

> A general discussion of methods of approximation, with extensive lists and numerical comparisons of approximations for the main continuous and discrete forms.

Molenaar, W. (1970), *Approximations to the Poisson, Binomial and Hypergeometric Distribution Functions*. Amsterdam: Mathematical Centre. (8.2, 8.4, 8.5)

> Approximations for these distributions are discussed in a unified way, and "preferred" forms chosen on the basis of extensive numerical comparisons.

Patil, G.P. (ed.) (1965), *Classical and Contagious Discrete Distributions*. New York: Pergamon.

> Conference proceedings (McGill, 1963). Papers drawn from this volume are referred to in Section II, *below*, as *"Discrete Distributions"*.

Patil, G.P. (1970), *Random Counts in Scientific Work* (3 vol.). Univ. Park, Penn.: State Univ. Press.

> Conference proceedings (Dallas, 1968) amplified by invited papers. Papers drawn from this volume are referred to in Section II, *below*, as *"Random Counts"*.

Patil, G.P. and Joshi, S.W. (1968), *A Dictionary and Bibliography of Discrete Distributions*. Edinburgh: Oliver and Boyd. (ex. 5.19, 6.1)

> Gives a list of all the discrete distributions and their properties, with over 3000 references. These references are then coded by type of distribution, type of statistical inference, and field of application. Subsequent users of each paper are also listed.

211

II REFERENCES AND AUTHOR INDEX

Adelson, R.M. (1966a), Compound Poisson distributions, *Opl. Res. Q.*, 17, 73—6. (4.1)

Adelson, R.M. (1966b), Reply to G.J. Goodhardt, *Opl. Res. Q.*, 17, 192—3. (4.1)

Ahuja, J.C. and Nash, S.W. (1967), The generalised Gompertz—Verhulst family of distributions, *Sankhyā, A*, 29, 141—56. (ex. 1.12)

Aitchison, J. (1955), On the distribution of a positive random variable having a discrete probability mass at the origin, *J. Amer. Statist. Ass.*, 50, 901—8. (4.11)

Aitchison, J. and Brown, J.A.C. (1957), *The Lognormal Distribution*, Cambridge: Cambridge University Press. (2.11, 2.12, 2.13, 3.6)

Aitken, A.C. and Gonin, A.T. (1935), On fourfold sampling with and without replacement, *Proc. Roy. Soc. Edin.*, 55, 114—25. (7.11)

Anscombe, F.J. (1948), The transformation of Poisson, binomial and negative binomial data, *Biometrika*, 35, 246—54. (8.8, 8.9)

Anscombe, F.J. (1949), Large sample theory of sequential estimation, *Biometrika*, 36, 455—8. (5.3)

Anscombe, F.J. (1950), Sampling theory of the negative binomial and logarithmic series distributions, *Biometrika*, 37, 358—82. (ex. 5.11, 6.7, 6.10, ex. 6.8)

Arbous, A.G. and Kerrich, J.C. (1951), Accident statistics and the concept of accident proneness, *Biometrics*, 7, 340—432. (6.6)

Aroian, L.A. (1937), The type B Gram—Charlier series, *Ann. Math. Statist.*, 8, 183—92. (5.13)

Atkinson, A.C. (1970), A method for discriminating between models, *J. Roy. Statist. Soc., Ser. B*, 32, 323—53. (1.6).

Barankin, E.W. and Gurland J. (1951), On asymptotically efficient normal estimators, *Univ. Calif. Publ. in Statist.*, 1, 89—129. (6.11)

Bardwell, G.E. and Crow, E.L. (1964), A two parameter family of hyper-Poisson distributions, *J. Amer. Statist. Ass.*, 59, 133—41. (5.3, 6.4, 6.14)

Bardwell, G.E. and Crow, E.L. (1965), Estimation of parameters of the hyper-Poisson distributions, *Discrete Distributions*, 127—40. (5.3, 6.4, 6.14)

Barndoff-Neilsen, O. (1965), Identifiability of mixtures of exponential families, *J. Math. Anal. Appl.*, 12, 115—21. (4.5)

Bartholomew, D.J. (1969), Sufficient condition for a mixture of exponentials to be a probability density function, *Ann. Math. Statist.*, 40, 2183—8. (ex. 4.6)

Bartko, J.J. (1961), The negative binomial distribution: a review of properties and applications, *Virg. J. Sci. (NS)*, 12, 18—37. (5.3)

Bartlett, M.S. (1936), Square root transformations in the analysis of variance, *J. Roy. Statist. Soc., Ser. B*, 3, 68—78. (8.9)

Bartlett, M.S. (1947), The use of transformations, *Biometrics*, 3, 39—52. (8.8)

Bartlett, M.S. and Kendall, D.G. (1946), Statistical analysis of variance heterogeneity and the logarithmic transformation, *J. Roy. Statist. Soc., Ser. B*, 8, 128—38. (8.9)

Barton, D.E. (1957), The modality of Neyman's contagious distribution of type A, *Trab. Estadist.*, 8, 13—22. (6.7)

Barton, D.E. and Dennis, K.E. (1952), The conditions under which the Gram—Charlier and Edgeworth curves are positive definite and unimodal, *Biometrika*, 39, 425—7. (2.6)

Bates, G.E. (1955), Joint distributions of time intervals for the occurrence of successive accidents in a generalised Pólya scheme, *Ann. Math. Statist.*, 26, 705—20. (7.8)

Bates, G.E. and Neyman, J. (1952a, b), Contributions to the theory of accident proneness. I: An optimistic model of the correlation between light and severe accidents. II: True or false contagion, *Univ. Calif. Publ. in Statist.*, 1, 215–53, 255–75. (7.8)

Beale, E.M.L. and Mallows, C.L. (1959), Scale mixing of symmetric distributions with zero means, *Ann. Math. Statist.*, 30, 1145–51. (4.11)

Beale, F.S. (1941), On a certain class of orthogonal polynomials, *Ann. Math. Statist.*, 12, 97–103. (5.13)

Beall, G. (1940), The fit and significance of contagious distributions when applied to observations on larval insects, *Ecology*, 21, 460–74. (6.14)

Beall, G. (1942), The transformation of data from entomological field experiments so that the Analysis of Variance becomes applicable, *Biometrika*, 32, 243–62. (8.8, 8.9)

Beall, G. and Rescia, R.R. (1953), A generalisation of Neyman's contagious distributions, *Biometrics*, 9, 354–86. (6.6, ex. 6.2)

Bergström, H. (1952), On some expansions of stable distributions, *Arkiv för Mathematik*, 2, 375–8. (ex. 2.8)

Bhat, B.R. and Kulkarni, N.V. (1966), On efficient multinomial estimation, *J. Roy. Statist. Soc., Ser. B.*, 28, 45–52. (7.6)

Bhattacharya, S.K. (1966), Confluent hypergeometric distributions of discrete and continuous type with applications to accident proneness, *Calcutta Statist. Ass. Bull.*, 15, 20–51. (6.4)

Bhattacharya, S.K. (1967), A result on accident proneness, *Biometrika*, 54, 324–5. (7.8)

Bhattacharya, S.K. and Holla, M.S. (1965), On a discrete distribution with special reference to the theory of accident proneness, *J. Amer. Statist. Ass.*, 60, 1060–6. (ex. 6.15, 7.7)

Binet, F.E. (1968), Generalisation of the hypergeometric distribution, *Magyar Tud. Akad. Math. Fiz. Ozst. Kösl.*, 18, 137–46 (in Hung.). (5.1)

Birnbaum, A. and Laska, E. (1967), Optimal robustness: a general method with applications to linear estimators of location, *J. Amer. Statist. Ass.*, 62, 1230–41. (1.9)

Bissinger, B.H. (1965), A type-resisting distribution generated from considerations of an inventory decision model, *Discrete Distributions*, 15–17. (ex. 5.19)

Blischke, W.R. (1964), Estimating the parameters of mixtures of binomial distributions, *J. Amer. Statist. Ass.*, 59, 510–28. (4.5)

Blischke, W.R. (1965), Mixtures of discrete distributions, *Discrete Distributions*, 351–72. (4.3, 4.5)

Bliss, C.I. and Fisher, R.A. (1953), Fitting the negative binomial distribution to biological data; the efficient fitting of the negative binomial, *Biometrics*, 9, 176–200. (6.12, 8.8)

Blom, G. (1954), Transformations of the binomial, negative binomial, Poisson and χ^2 distributions, *Biometrika*, 41, 302–16. (8.9)

Boas, R.P. (1949), Representation of probability distributions by Charlier series, *Ann. Math. Statist.*, 20, 376–92. (2.5, 5.13).

Boes, D.C. (1966), On the estimation of mixing distributions, *Ann. Math. Statist.*, 37, 177–88. (4.5)

Bolshev, L.N., Gladkov, B.V., and Sheneglova, M.V. (1961), Tables for the calculation of B- and z-distribution functions, *Theor. Prob. & Appl.*, 6, 410–19. (8.4, 8.5)

Borges, R. (1970), Eine Approximation der Binomialverteilung durch die Normalverteilung der Ordnung $1/n$, *Zeit. Wahrsch. u. verw. Gebiete*, 14, 189–99. (8.5)

Bosch, A.J. (1963), The Pólya distribution, *Stat. Neer.*, 17, 201–13. (5.4)

214

Boswell, M.T. and Patil, G.P. (1970), Chance mechanisms generating the negative binomial distribution, *Random Counts*, Vol. 1, 3–22. (5.3)

Bowman, K. and Shenton, L.R. (1970), Properties of the ML estimator for the parameter of the logarithmic series distribution, *Random Counts*, Vol. 1, 127–50. (5.11)

Box, G.E.P. and Cox, D.R. (1964), An analysis of transformations, *J. Roy. Statist. Soc., Ser. B*, 26, 211–52. (8.8)

Boyd, A.V. (1959), Inequalities for Mills' ratio, *Rep. Statist. Appl. Res. Un. Jap. Sci. Engrs.*, 6, 44–6. (8.5)

Brass, W. (1958), Simplified methods of fitting the truncated negative binomial distribution, *Biometrika*, 45, 59–68. (5.11)

Burr, I.W. (1942), Cumulative frequency functions, *Ann. Math. Statist.*, 13, 215–32. (1.1, 2.6)

Burr I.W. (1952). Formulae for approximating the hypergeometric and the binomial by the Poisson distribution, *Ann. Math. Statist.*, 23, 145 (abstract). (8.5)

Burr, I.W. (1967), A useful approximation to the normal distribution function with an application to simulation, *Technometrics*, 9, 647–51. (ex. 2.6)

Burr, I.W. (1968), On a general system of distributions, III, *J. Amer. Statist. Ass.*, 63, 636–43. (2.16)

Burr, I.W. and Cislak, P.J. (1968), On a general system of distributions, I, II, *J. Amer. Statist. Ass.*, 63, 627–35. (2.16)

Buslik, D. (1950), Mixing and sampling with special reference to multi-sized granular material, *Bull. Amer. Soc. Testing Materials*, 66, 165. (4.11)

Camp, B.H. (1951), Approximation to the point binomial, *Ann. Math. Statist.*, 22, 130–1. (8.5)

Carver, H.C. (1919), On the graduation of frequency distributions, *Proc. Casualty Actuarial Soc. Amer.*, 6, 52–72. (2.4, 5.1)

Carver, H.C. (1923), Chapter VII on Frequency Curves in *Handbook of Mathematical Statistics*, ed. H.L. Rietz, Cambridge, Mass.: Riverside. 92–119. (2.4)

Cassie, R. (1962), Frequency distribution models in the ecology of plankton and other organisms, *J. Animal Ecol.*, 31, 65–92. (4.8)

Cernuischi, F. and Castagnetto, L. (1946), Chains of rare events, *Ann. Math. Statist.*, 17, 51–61. (5.4)

Chambers, J.M. (1967), On methods of asymptotic approximation for multivariate distributions, *Biometrika*, 54, 367–83. (3.5, 8.7)

Charlier, C.V.L. (1905), Ueber die Darstellung willkürlicher Funktionen, *Ark. f. Math. Astr. o. Fys.*, 2, 1–35. (5.13)

Charlier, C.V.L. (1907), Die Zweite Form des Fehlergesetzes, *Ark. f. Math. Astr. o. Fys.*, 2, No. 15, 1–8. (2.1)

Charlier, C.V.L. (1928), A new form of the frequency function, *Medd. Lunds Astr. Obs.*, 51, 1–28. (2.7)

Chernoff, H. and Rubin, M. (1955), The estimation of the location of a discontinuity in density, *Proc. 3rd Berk. Symp.*, 1, 19–37. (1.5)

Choi, K. and Bulgren, W.G. (1968), An estimation procedure for mixtures of distributions, *J. Roy. Statist. Soc., Ser. B.*, 30, 444–60. (4.5)

Clenshaw, W.C., Miller, G.F. and Woodger, M. (1962), Algorithms for special functions, *Num. Math.*, 4, 403–19. (8.5)

Cliff, A.D. and Ord, J.K. (1969), The problem of spatial autocorrelation, in *Studies in Regional Science*, ed. A.J. Scott, London: Pion. 25–55. (6.15)

Cliff, A.D. and Ord, J.K. (1970), Spatial autocorrelation: a review of existing and new measures with applications, *Econ. Geogr.*, 46, 269–92. (6.15)

Cochran, W.G. (1940), The analysis of variance when experimental errors follow the Poisson or binomial laws, *Ann. Math. Statist.*, 11, 335–47. (8.8)

Cochran, W.G. (1943), Analysis of variance for percentages based on unequal numbers, *J. Amer. Statist. Ass.*, 38, 287–301.   (8.8)

Cohen, A.C., Jr. (1951), Estimating the parameters of the logarithmic normal distribution by maximum likelihood, *J. Amer. Statist. Ass.*, 46, 206–12. (2.13)

Cohen, A.C., Jr. (1954), Estimation of the Poisson parameter from truncated samples and from censored samples, *J. Amer. Statist. Ass.*, 49, 158–168.   (5.11)

Cohen, A.C., Jr. (1960a), Estimating the parameter in a conditional Poisson distribution, *Biometrics*, 16, 203–11.   (4.11, 5.11)

Cohen, A.C., Jr. (1960b), An extension of a truncated Poisson distribution, *Biometrics*, 16, 446–50.   (5.11)

Cohen, A.C., Jr. (1960c), Estimating the parameters of a modified Poisson distribution, *J. Amer. Statist. Ass.*, 55, 139–43.   (5.11)

Cohen, A.C., Jr. (1960d), Estimation in truncated Poisson distributions when zeros and some ones are missing, *J. Amer. Statist. Ass.*, 55, 342–48. (5.11)

Cohen, A.C., Jr. (1961), Estimating the Poisson distribution from samples that are truncated on the right, *Technometrics*, 3, 433–38.   (5.11, ex. 5.16)

Cohen, A.C., Jr. (1967), Estimation in mixtures of two normal distributions, *Technometrics*, 9, 15–28.   (4.8)

Cornish, E.A. and Fisher, R.A. (1937), Moments and cumulants in the specification of distributions, *Rev. Int. Statist. Inst.*, 14, 1–14.   (2.8)

Cox, D.R. (1970), *The Analysis of Binary Data*. London: Methuen.   (8.8)

Cramér, H.C. (1928), On the composition of elementary errors, *Skand. Aktuar.*, 11, 13–74, 141–80.   (2.3)

Cresswell, W.L. and Froggart, P. (1963), *The causation of bus driver accidents— an epidemiological study*, London: Oxford Univ. Press.   (6.6)

Crow, E.L. and Siddiqui, M.M. (1967), Robust estimation of location, *J. Amer. Statist. Ass.*, 62, 353–89.   (1.9)

Daniels, H.E. (1944), Measures of correlation in the universe of sample permutations, *Biometrika*, 33, 129–35.   (3.9)

Daniels, H.E. (1954), Saddlepoint approximations in statistics, *Ann. Math. Statist.*, 25, 631–50.   (2.6)

Daniels, H.E. (1961), Mixtures of geometric distributions, *J. Roy. Statist. Soc.*, Ser. B., 23, 409–13.   (4.5, 4.11)

Darwin, J.H. (1957), The power of the Poisson index of dispersion, *Biometrika*, 44, 286–9.   (6.9)

David, F.N. and Johnson, N.L. (1952), The truncated Poisson distribution, *Biometrics*, 8, 275–85.   (5.11)

Davies, O.L. (1933, 1934), On asymptotic formulae for the hypergeometric series, *Biometrika*, 25, 295–322 (Part I).   (5.1, 5.7), and 26, 59–107 (Part II).   (5.1, 5.12).

Day, N.E. (1969), Estimating the components of a mixture of normal distributions, *Biometrika*, 56, 463–74.   (4.8)

De Bruijn, N.G. (1961), *Asymptotic Methods in Analysis*, New York: Wiley. (8.2)

Deely, J.J. and Kruse, R.L. (1968), Construction of sequences estimating the mixing distribution, *Ann. Math. Statist.*, 39, 286–8.   (4.5)

De Moivre, A. (1718), *The Doctrine of Chances*.   (8.1)

Doss, S.A.D.C. (1969), Characterisations of the linear exponential family in a parameter by recurrence relations for functions of the cumulants, *Ann. Math. Statist.*, 40, 1721–7.   (1.12)

Douglas, J.B. (1955), Fitting the Neyman Type A (two parameter) contagious distributions, *Biometrics*, 11, 149–73.   (6.12)

216

Douglas, J.B. (1965), Asymptotic expansions for some contagious distributions, *Discrete Distributions*, 291–302. (8.6)

Dubey, S.D. (1966), Graphical tests for discrete distributions, *Amer. Statistician*, 20, 23–5. (5.10)

Dubey, S.D. (1968), A compound Weibull distribution, *Naval Res. Logist. Q.*, 15, 179–88. (ex. 4.11)

Eagleson, G.K. and Lancaster, H.O. (1967), The regression system of sums with random elements in common, *Austr. J. Statist.*, 9, 119–25. (3.7)

Edgeworth, F.Y. (1896), The asymmetric probability curve, *Phil. Mag.*, 41, 90–99. (2.2)

Edgeworth, F.Y. (1898), On the representation of statistics by mathematical formulae, *J. Roy. Statist. Soc., Ser. A.*, 61, 670–700. (1.1, 2.11)

Edwards, C.B. and Gurland, J. (1961), A class of distributions applicable to accidents, *J. Amer. Statist. Ass.*, 56, 503–17.

Ehrenberg, A.S.C. (1959), The pattern of consumer purchases, *Appl. Statist.*, 8, 26–41. (5.10)

Eisenhart, C. (1947), The assumptions underlying the analysis of variance, *Biometrics*, 3, 1–21. (8.8)

Erdelyi, A. *et al.* (1953), *Higher Transcendental Functions*, Vol. 2, New York: McGraw–Hill. (App. A)

Erdelyi, A. (1956), *Asymptotic Expansions*, New York: Dover. (8.2)

Esseen, C.G. (1945), Fourier analysis of distribution functions. A mathematical study of the Laplace–Gauss law, *Acta Math.*, 77, 1–125. (8.4, 8.5)

Fama, E.F. and Roll, R. (1968), Some properties of symmetric stable distributions, *J. Amer. Statist. Ass.*, 63, 817–36. (2.8)

Farlie, D.J.G. (1960), University of London Ph.D. thesis (unpublished). (3.8, 3.9)

Fechner, G.T. (1897), *Kollektivmasslehre*. Leipzig: Engelmann. (1.1)

Feller, W. (1943), On a general class of contagious distributions, *Ann. Math. Statist.*, 14, 389–400. (4.1, ex. 4.2, 6.6)

Feller, W. (1957, 1966), *An Introduction to Probability Theory and its Applications*, Vol. I and II. New York: Wiley. (I: 4.1, 7.2; II: 2.3, 3.5)

Fellingham, S.A. and Stoker, D.J. (1964), An approximation for the exact distribution of the Wilcoxon test for symmetry, *J. Amer. Statist. Ass.*, 59, 899–905. (2.10, ex. 2.3)

Finney, D.J. (1941), On the distribution of a variate whose logarithm is normally distributed, *J. Roy. Statist. Soc., Ser. B*, 7, 155–61. (8.9)

Finucan, H.M. (1964), The mode of a multinomial distribution, *Biometrika*, 51, 513–17. (5.8, 7.4)

Fisher, R.A. (1921a), On the mathematical foundations of theoretical statistics, *Phil. Trans. Roy. Soc., A*, 222, 309–68. (1.12)

Fisher, R.A. (1921b), On the probable error of a coefficient of correlation deduced from a small sample, *Metron*, 1, 3–32. (2.11)

Fisher, R.A. (1937), Professor Karl Pearson and the method of moments, *Ann. Eugen.*, 7, 303–18. (1.6)

Fisher, R.A. (1941), The negative binomial distribution, *Ann. Eugen.* 11, 182–7. (6.6)

Fisher, R.A. (1953), see Bliss and Fisher (1953).

Fisher, R.A. and Cornish, E.A. (1960), The percentile points of distributions having known cumulants, *Technometrics*, 2, 209–25. (2.8)

Fréchet, M. (1951), Sur les tableaux de corrélation dont les marges sont donnees, *Ann. Univ. Lyon, A*, Sér. 3, 14, 53–77. (3.8)

Freedman, D.A. (1965), Bernard Friedman's urn, *Ann. Math. Statist.*, 36, 356–370. (5.4)

Freeman, M.F. and Tukey, J.W. (1950), Transformations related to the angular and the square root, *Ann. Math. Statist.*, 21, 607—11. (8.9)

Frisch, R. (1932), On the use of difference equations in the study of frequency distributions, *Metron*, 10, No. 3, 35—59. (5.1, 5.6)

Gabriel, K. (1959), The distribution of the number of successes in a sequence of dependent trials, *Biometrika*, 46, 454—60. (5.3)

Galton, F. (1879), The geometric mean in vital and social statistics, *Proc. Roy. Soc.*, 29, 365—6. (1.1, 2.11)

Gart, J.J. (1970), Some simple graphically oriented statistical methods for discrete data, *Random Counts*, Vol.I, 171—91. (5.10)

Gebhardt, F. (1969), Some numerical comparisons of several approximations to the binomial distribution, *J. Amer. Statist. Ass.*, 64, 1638—46. (8.5)

Gnedenko, B.V. and Kolmogorov, A.N. (1954), *Limit Distributions for Sums of Independent Random Variables*, Cambridge, Mass.: Addison-Wesley. (8.5)

Gonin, H.T. (1961), The use of orthogonal polynomials of the positive and negative binomial frequency functions in curve fitting by Aitken's method, *Biometrika*, 48, 115—23. (5.13)

Goodhardt, G.J. (1966), A letter on "Poisson distributions", *Opl. Res. Q.*, 17, 191—2. (4.1)

Govindarajulu, Z. (1962), The reciprocal of the decapitated negative binomial distribution, *J. Amer. Statist. Ass.*, 57, 906—13. (5.11)

Govindarajulu, Z. (1963), Recurrence relations for the inverse moments of the positive binomial variate, *J. Amer. Statist. Ass.*, 58, 468—73. (5.11)

Govindarajulu, Z. (1964), The first two moments of the reciprocal of the positive hypergeometric variable, *Sankhyā*, Ser. B, 26, 217—36. (5.11)

Govindarajulu, Z. (1965), Normal approximations to classical discrete distributions, *Discrete Distributions*, 79—108. (8.5, 8.6)

Grab, E.L. and Savage, I.R. (1954), Tables of the expected value of $1/X$ for positive Bernoulli and Poisson variables, *J. Amer. Statist. Ass.*, 49, 169—77. (5.11)

Greenleaf, H.E.H. (1932), Curve approximation by means of functions analogous to the Hermite polynomials, *Ann. Math. Statist.*, 3, 204—55. (5.13)

Greenwood, M. and Yule, G.U. (1920), An enquiry into the nature of frequency distributions representative of multiple happenings, *J. Roy. Statist. Soc.*, Ser. A, 83, 255—79. (6.6)

Guldberg, A. (1931), On discontinuous frequency functions and statistical series, *Skand. Aktuar.*, 14, 167—87. (5.1, 5.6)

Gumbel, E.J. (1960), The bivariate exponential distribution, *J. Amer. Statist. Ass.*, 55, 698—708. (3.8)

Gumbel, E.J. (1961), The bivariate logistic distribution, *J. Amer. Statist. Ass.*, 56, 335—49. (3.8)

Gumbel, E.J. and Mustafi, C.K. (1967), Some analytical properties of bivariate extremal distributions, *J. Amer. Statist. Ass.*, 62, 569—88. (3.8)

Gurland, J. (1957), Some interrelations among compound and generalised distributions, *Biometrika*, 44, 265—8. (4.14)

Gurland, J. (1958), A generalised class of contagious distributions, *Biometrika*, 14, 229—49. (6.6, ex. 6.2, ex. 6.3)

Gurland, J. (1965), A method of estimation for some generalised Poisson distributions, *Discrete Distributions*, 141—58. (6.7, 6.11)

Haight, F.A. (1961), A distribution analogous to the Borel—Tanner, *Biometrika*, 48, 167—73. (5.12)

Haight, F.A. (1967), *Handbook of the Poisson distribution*. Publications in O.R. No. 11, New York: Wiley. (5.3, 5.11, ex. 5.16, 6.1)

Haldane, J.B.S. (1937), The approximate normalisation of a class of frequency distributions, *Biometrika*, 39, 392—402. (8.5, 8.9)

218

Haldane, J.B.S. (1942), Mode and median of a nearly normal distribution with given moments, *Biometrika*, **32**, 294—9.   (ex. 2.5)

Hall, W.J. (1956), Some hypergeometric series distributions occurring in birth and death processes at equilibrium (preliminary report), *Ann. Math. Statist.*, **27**, 221.   (6.4)

Hansmann, G.H. (1934), On certain non-normal symmetrical frequency distributions, *Biometrika*, **26**, 129—95.   (1.5)

Harding, J.P. (1949), The use of probability paper for the graphical analysis of polymodal frequency distributions, *J. Marine Biol. Ass.*, **28**, 141—52. (4.8, ex. 4.9, ex. 4.10)

Hart, R.G. (1957), A formula for the approximation of definite integrals of the normal distribution functions, *Math. Tables Aids Comput.*, **11**, 265. (8.5).

Hartley, H.O. (1958), Maximum likelihood estimation from incomplete data, *Biometrics*, **14**, 174—94.   (5.11)

Hassanein, K.M. (1969), Estimation of the parameters of the extreme value distribution by use of two or three order statistics, *Biometrika*, **56**, 429—36.   (2.16)

Hasselblad, V. (1969), Estimation of finite mixtures of distributions from the exponential family, *J. Amer. Statist. Ass.*, **64**, 1459—71.   (4.7)

Hastings, C. (1957), *Approximations for Digital Computers*. New York: Princeton University.   (8.3, 8.5)

Hatke, A.M. (1949), A certain cumulative probability function, *Ann. Math. Statist.*, **20**, 461—3.   (2.16)

Henderson, J. (1922), On expansions in tetrachoric series, *Biometrika*, **14**, 157—85.   (2.3)

Hildebrandt, E.H. (1931), Systems of polynomials connected with the Charlier expansions and the Pearson differential and difference equations, *Ann. Math. Statist.*, **2**, 379—439.   (5.13)

Hinz, P. and Gurland, J. (1967), Simplified techniques for estimating parameters of some generalised Poisson distributions, *Biometrika*, **54**, 555—66.   (6.11, 6.14)

Hinz, P. and Gurland, J. (1968), A method of analysing untransformed data from the negative binomial and other contagious distributions, *Biometrika*, **55**, 163—70.   (8.8)

Hinz, P. and Gurland, J. (1970), Tests of fit for the negative binomial and other contagious distributions, *J. Amer. Statist. Ass.*, **65**, 887—903.

Holgate, P. (1964), Estimation of the bivariate Poisson, *Biometrika*, **51**, 241—5.   (7.5, ex. 7.5)

Holgate, P. (1966), Bivariate generalisations of Neyman's Type A distributions, *Biometrika*, **53**, 241—5.   (7.7)

Holgate, P. (1970), The modality of some compound Poisson distributions, *Biometrika*, **57**, 666—7.   (6.7)

Hotelling, H. and Frankel, L.R. (1938), The transformation of statistics to simplify their distribution, *Ann. Math. Statist.*, **9**, 87—96.   (8.4)

Hoyle, M.H. (1968), The estimation of variances after using a Gaussianating transformation, *Ann. Math. Statist.*, **39**, 1125—43.   (8.9)

Irwin, J.O. (1964), The personal factor in accidents — a review article. *J. Roy. Statist. Soc., Ser. A.*, **127**, 438—51.   (6.6)

Irwin, J.O. (1965), Inverse factorial series as frequency distributions, *Discrete Distributions*, 159—74.   (5.3)

Irwin, J.O. (1968), The generalised Waring distribution applied to accident data, *J. Roy. Statist. Soc., Ser. A.*, **131**, 205—25.   (5.3)

Ishii, G. and Hayakawa, R. (1960), On the compound binomial distribution, *Ann. Inst. Statist. Math.*, **12**, 69—80.   (7.3, ex. 7.2)

Isserlis, L. (1914), The application of solid hypergeometrical series to frequency distributions in space, *Phil. Mag.*, **28**, 379—403.   (3.2)

Janardan, K.G. and Patil, G.P. (1970), Location of modes for certain univariate and multivariate distributions, *Random Counts*, Vol. 1, 57–75.   (5.6)

Jenkins, G.M. and Watts, D.G. (1968), *Spectral Analysis and its Applications*, San Francisco: Holden-Day.   (8.9)

Johnson, N.L. (1949a), Systems of frequency curves generated by methods of translation, *Biometrika*, 36, 149–76.   (1.1, 2.12, 2.13, 2.15, 2.19)

Johnson, N.L. (1949b), Bivariate distributions based on simple translation systems. *Biometrika*, 36, 297–304.   (1.1, 2.12, 3.7, ex. 3.3)

Johnson, N.L. (1957), A note on the mean deviation of the binomial distribution, *Biometrika*, 44, 532.   (7.8)

Johnson, N.L. (1965), Tables to facilitate fitting S_U frequency curves, *Biometrika*, 52, 547–58.   (2.15)

Johnson, N.L. *et al.* (1963), Tables of percentage points of Pearson curves, for given $\sqrt{\beta_1}$ and β_2 expressed in standard measure, *Biometrika*, 50, 459–98.   (1.8)

Joshi, S.W. and Patil, G.P. (1970), A class of statistical models for multiple counts, *Random Counts*, Vol. 1, 189–203.   (7.4, 7.6)

Kabir, A.B.M.L. (1968), Estimation of parameters of a finite mixture of distributions, *J. Roy. Statist. Soc., Ser. B.*, 30, 472–82.   (4.7)

Kalbfleisch, J.D. and Sprott, D.A. (1970), Application of likelihood methods to models involving large numbers of parameters, *J. Roy. Statist. Soc., Ser. B.*, 32, 175–208.   (7.5)

Kamat, A.R. (1965), Incomplete and absolute moments of some discrete distributions, *Discrete Distributions*, 45–64.   (5.6)

Kamat, A.R. (1966), A generalisation of Johnson's property of the mean deviation for a class of discrete distributions, *Biometrika*, 53, 285–7.   (5.6)

Kapteyn, J.C. (1903), *Skew frequency curves in biology and statistics*, Astronomical Laboratory, Groningen: Noordhoff.   (1.1, 2.11)

Katti, S.K. (1960), Moments of the absolute difference and absolute deviation of discrete distributions, *Ann. Math. Statist.*, 31, 78–85.   (5.6)

Katti, S.K. and Gurland, J. (1961), The Poisson Pascal distribution, *Biometrics*, 17, 527–38.   (4.1, 6.6, 6.12)

Katti, S.K. and Gurland, J. (1962), Some methods of estimation for the Poisson–binomial distribution, *Biometrics*, 18, 42–51.   (6.12)

Katti, S.K. and Sly, L.E. (1965), Analysis of contagious data through behaviouristic models, *Discrete Distributions*, 303–19.   (6.7, 7.6)

Katz, L. (1946), On the class of distributions defined by the difference equation $(x + 1)f(x + 1) = (a + bx)f(x)$ (abstract), *Ann. Math. Statist.*, 17, 501.   (5.1, 5.3)

Katz, L. (1948), Frequency functions defined by the Pearson difference equation (abstract), *Ann. Math. Statist.*, 19, 120.   (5.1)

Katz, L. (1965), Unified treatment of a broad class of discrete probability distributions, *Discrete Distributions*, 175–82.   (5.1)

Kemp, C.D. (1967a), On a contagious distribution suggested for accident data, *Biometrics*, 23, 241–55.   (6.6, ex. 6.17)

Kemp, C.D. (1967b), "Stuttering Poisson" distributions, *J. Statist. & Social Inq. Soc. Ireland*, 21, 151–7.   (6.1)

Kemp, C.D. (1970), Accident proneness and discrete distribution theory, *Random Counts*, Vol. 2, 41–65.   (6.6)

Kemp, C.D. and Kemp, A.W. (1956), Generalised hypergeometric distributions, *J. Roy. Statist. Soc., Ser. B*, 18, 202–11.   (5.1)

Kemp, C.D. and Kemp, A.W. (1965), Some properties of the Hermite distribution, *Biometrika*, 52, 381–94.   (6.6)

Kemp, C.D. and Kemp, A.W. (1966), An alternative derivation of the Hermite distribution, *Biometrika*, 53, 627–8.   (6.5)

220

Kemp, C.D. and Kemp, A.W. (1969), Some distributions arising from an inventory decision problem, *Bull. Int. Statist. Inst.*, **43**, Bk. 2, 367–9. (ex. 5.20)

Kemsley, W.F.F. (1952), Body weight at different ages and heights, *Ann. Eugen.* **16**, 316–34. (2.13)

Kendall, D.G. (1948a), On some models of population growth leading to R.A. Fisher's logarithmic distribution, *Biometrika*, **35**, 6–15. (5.4)

Kendall, D.G. (1948b), On the generalised "birth and death" process, *Ann. Math. Statist.*, **19**, 1–15. (5.4)

Kendall, M.G. (1941), A recurrence relation for the semi-invariants of Pearson curves, *Biometrika*, **32**, 81–2. (1.4, ex. 1.1)

Kendall, M.G. (1949), Rank and product moment correlation, *Biometrika*, **36**, 177–93. (3.4)

Kendall, M.G. (1961), Natural law in the social sciences, *J. Roy. Statist. Soc.*, Ser. *A*, **124**, 1–18. (ex. 5.17)

Kendall, M.G. and Buckland, W.R. (1960), *A Dictionary of Statistical Terms*. London: Oliver & Boyd. (4.1)

Kendall, M.G. and Stuart, A. (1950), The law of the cubic proportion in election results, *Brit. J. Sociology*, **1**, 183–96. (2.16)

Kendall, M.G. and Stuart A. (1969, 1967), *The Advanced Theory of Statistics*, Vol. I and II. London: Griffin. (I: 1.3, 2.2, 2.3, 2.8, 8.3; II: 1.12)

Khatri, C.G. (1959), On certain properties of power series distributions, *Biometrika*, **46**, 486–90. (6.2)

Khatri, C.G. and Patel, I.R. (1961), Three classes of univariate discrete distributions, *Biometrics*, **17**, 567–75. (6.8, ex. 6.10, ex. 6.12)

Khovanskii, A. (1963), *The Application of Continued Fractions and their Generalisations to Problems in Approximation Theory*. Amsterdam: Noordhoff. (8.3)

Knott, M. (1967), Sampling mixtures of particles, *Technometrics*, **9**, 365–71. (4.11)

Kolmogorov, A.N. (1933), *Grundbegriffe der Wahrscheinlichkeitsrechnung*. Berlin: Springer. (4.2)

Kosambi, D.D. (1949), Characteristic properties of series distributions, *Proc. Nat. Inst. Sci. India*, **15**, 109–13. (1.12)

Koshal, R.S. (1933), Application of the method of maximum likelihood to the improvement of curves fitted by the method of moments, *J. Roy. Statist. Soc., Ser. A.*, **96**, 303–13. (1.6)

Krishnamoorthy, A.S. (1951), Multivariate binomial and Poisson distributions, *Sankhyā*, **11**, 117–24. (7.3, 7.11)

Krishnamoorthy, A.S. and Parthasarthy, M. (1951), A multivariate gamma type distribution, *Ann. Math. Statist.*, **11**, 549–57. (3.4)

Kullback, S.K. (1947), On the Charlier type B series, *Ann. Math. Statist.*, **17**, 574–81. (5.13)

Lancaster, H.O. (1958), The structure of bivariate distributions, *Ann. Math. Statist.*, **29**, 719–36. (3.7, 7.1)

Laubscher, N.F. (1960), On the stabilisation of the Poisson variance, *Trab. Estadist.*, **11**, 199–207. (8.9)

Laubscher, N.F. (1961), On stabilising the binomial and negative binomial variances, *J. Amer. Statist. Ass.*, **56**, 143–50. (8.8)

Lehmann, E.L. (1959), *Testing Statistical Hypotheses*, New York: Wiley. (1.11)

Lindley, D.V. (1950), Grouping corrections and maximum likelihood equations, *Proc. Camb. Phil. Soc.*, **46**, 106–10. (1.6)

McAlister, D. (1879), The law of the geometric mean, *Proc. Roy. Soc.*, **29**, 367–76. (1.1, 2.11)

McFadden, J.A. (1955), On models of correlation and a comparison with the multivariate normal integral, *Ann. Math. Statist.*, **26**, 478–89. (ex. 5.5)

McGilchrist, C.A. (1969), Discrete distribution estimators from the recurrence equation for probabilities, *J. Amer. Statist. Ass.*, 64, 602—9. (5.10)

McGuire, J.U., Brindley, T.A. and Bancroft, T.A. (1957), The distribution of European corn-borer larvae in field-corn, *Biometrics*, 13, 65—78, and correction, 14, 432—4. (6.6, 6.14).

McKendrick, A.G. (1926), Applications of mathematics to medical problems, *Proc. Edin. Math. Soc.*, 44, 98—130. (5.11)

Maceda, E.C. (1948), On the compound and generalised Poisson distributions, *Ann. Math. Statist.*, 19, 414—16. (4.2)

Mahamunulu, D.M. (1967), A note on regression in the multivariate Poisson distribution, *J. Amer. Statist. Ass.*, 62, 251—8. (7.4)

Mardia, K.V. (1967), Some contributions to contingency-type bivariate distributions, *Biometrika*, 54, 235—49. (3.8, 3.10, ex. 3.7)

Marlow, W.H. (1965), On a class of factorial distributions, *Ann. Math. Statist.*, 36, 1066—8. (5.3)

Martin, D.C. and Katti, S.K. (1962), Approximations to the Neyman type A for practical problems, *Biometrics*, 18, 354—64. (8.6)

Martin, D.C. and Katti, S.K. (1965), Fitting of certain contagious distributions to some available data by the maximum likelihood method, *Biometrics*, 21, 34—48. (6.12)

Mauldon, J.G. (1959), A generalisation of the beta distribution, *Ann. Math. Statist.*, 30, 509—20. (3.2)

Mellinger, C.D., Gaffey, W.R., Sylvester, D.L. and Manheimer, D.I. (1965), A mathematical model with applications to a study of accident repeatedness among children, *J. Amer. Statist. Ass.*, 60, 1046—59. (7.10)

Merrell, M. (1933), On certain relationships between β_1 and β_2 for the point binomial, *Ann. Math. Statist.*, 4, 216—28. (ex. 5.7)

Mishra, C. (1955), University of London Ph.D. thesis (unpublished). (ex. 1.2)

Molenaar, W. (1965), Survey of estimation methods for a mixture of two normal distributions, *Stat. Neer.*, 19, 249—63. (4.8)

Molenaar, W. and Van Zwet, W.R. (1966), On mixtures of distributions, *Ann. Math. Statist.*, 37, 281—3. (4.2)

Moore, P.G. (1957), Transformations to normality using fractional powers of the variable, *J. Amer. Statist. Ass.*, 44, 174—212. (8.4)

Morgenstern, D. (1956), Einfache Beispiele zweidimensionaler Verteilungen, *Mitt. Math. Statist.*, 8, 234—5. (3.8)

Mosimann, J.E. (1962), On the compound multinomial distribution, the multivariate beta distribution, and correlations among proportions, *Biometrika*, 49, 64—82. (7.3, 7.5)

Mosimann, J.E. (1963), On the compound negative multinomial distribution and correlations among inversely sampled pollen counts, *Biometrika*, 50, 46—54. (7.3, 7.5)

Mouzon, E.D. (1930), Equimodal frequency distributions, *Ann. Math. Statist.*, 1, 137—58. (1.5)

Munro, A.H. and Wixley, R.A.J. (1970), Estimates based on order statistics of small samples from a three parameter lognormal distribution, *J. Amer. Statist. Ass.*, 65, 212—25. (2.13)

Narumi, S. (1923), On the general forms of bivariate frequency distribution, *Biometrika*, 15, 77—88 and 209—21. (3.3)

Newbold, E.M. (1927), Practical applications of the statistics of repeated events, particularly to industrial accidents. *J. Roy. Statist. Soc.*, Ser. A., 90, 487—518. (6.6)

Neyman, J. (1939), On a new class of "contagious distributions" applicable in entomology and bacteriology, *Ann. Math. Statist.*, 10, 35—57. (6.6, 6.12)

Neyman, J. and Scott, E.L. (1960), Correction for bias introduced by a transformation of variables, *Ann. Math. Statist.*, 31, 643—55. (8.9)

222

Nicholson, W.L. (1956), On the normal approximation to the hypergeometric distribution, *Ann. Math. Statist.*, **27**, 471—83. (8.5)

Noack, A. (1950), A class of random variables with discrete distributions, *Ann. Math. Statist.*, **21**, 127—32. (6.2)

Ord, J.K. (1967a), University of London Ph.D. thesis (unpublished). (5.1, 5.6, 5.7, ex. 5.18, 8.8)

Ord, J.K. (1967b), On a system of discrete distributions, *Biometrika*, **54**, 649—56. (1.7, 5.1, 5.7)

Ord, J.K. (1967c), Graphical methods for a class of discrete distributions, *J. Roy. Statist. Soc.*, Ser. A, **130**, 232—8. (5.9, ex. 5.13)

Ord, J.K. (1968a), Approximations to distribution functions which are hypergeometric series, *Biometrika*, **55**, 243—8. (8.4, ex. 8.2)

Ord, J.K. (1968b), The discrete Student's *t* distribution, *Ann. Math. Statist.*, **39**, 1513—16. (1.10, ex. 5.4, ex. 5.12)

Ord, J.K. (1970), The negative binomial model and quadrat sampling, *Random Counts*, Vol. 2, 151—63. (6.9, 6.15, ex. 6.14)

Pairman, E. and Pearson, K. (1919), On corrections for the moment coefficients of limited range frequency distributions when there are finite or infinite ordinates and any slopes at the terminals of the range, *Biometrika*, **12**, 231—58. (1.6)

Pandey, K.N. (1964/5), On the generalised inflated Poisson distribution, *J. Sci. Res. Banaras Hindu Univ.*, **15**, No. 2, 157—62. (4.11)

Patil, G.P. (1961), Asymptotic bias and efficiency of ratio estimates for GPSD and certain applications, *Sankhyā*, A, **23**, 269—80. (6.3)

Patil, G.P. (1962a), On certain properties of the generalised power series distributions, *Ann. Inst. Statist. Math.*, **14**, 179—82. (6.2)

Patil, G.P. (1962b), Maximum likelihood estimation for GPSD and its application to a truncated binomial distribution, *Biometrika*, **49**, 227—37. (ex. 5.16, 6.3)

Patil, G.P. (1962c), Estimation by the two moments method for GPSD with applications, *Sankhyā*, B, **24**, 201—14. (6.3)

Patil, G.P. (1963), A characterisation of the exponential type distribution, *Biometrika*, **50**, 205—7. (1.12)

Patil, G.P. (1964), On certain compound Poisson and compound binomial distributions, *Sankhyā*, A, **26**, 293—4. (6.6)

Patil, G.P. (1965a), On the multivariate GPSD and its application to the multinomial and negative multinomial, *Discrete Distributions*, 183—94. (3.11, 7.6, ex. 7.6)

Patil, G.P. (1965b), Certain characteristic properties of multivariate discrete probability distributions akin to the Bates—Neyman model of accident proneness, *Sankhyā*, A, **27**, 259—70. (7.9)

Patil, G.P. and Bildikar, S. (1966), Identifiability of countable mixtures of discrete probability distributions using methods of infinite matrices, *Proc. Camb. Phil. Soc.*, **62**, 485—94. (4.3)

Patil, G.P. and Bildikar, S. (1967), Multivariate logarithmic series distribution as a probability model in population and community ecology, and some of its statistical properties, *J. Amer. Statist. Ass.*, **62**, 655—74. (7.3, 7.7)

Patil, G.P. and Shorrock, R. (1965), On certain properties of the exponential-type families, *J. Roy. Statist. Soc.*, Ser. B, **27**, 94—9. (1.12)

Patil, G.P. and Wani, J.K. (1965), On certain properties of the L.S.D. and the first type Stirling distribution, *Sankhyā*, A, **27**, 271—80. (5.11)

Patil, G.P. and Wani, J.K. (1966), Minimum variance unbiased estimation of the distribution function admitting a sufficient statistic, *Ann. Inst. Statist. Math.*, **18**, 39—47. (6.1)

Paulson, E. (1942), An approximate normalisation of the analysis of variance distribution, *Ann. Math. Statist.*, **13**, 233—5. (8.5)

Pearson, E.S. (1925), Bayes' theorem, examined in the light of experimental sampling, *Biometrika*, **17**, 388–442. (ex. 5.7)

Pearson, E.S. (1948), *Karl Pearson's Early Statistical Papers* (incl. series of contributions to the Math. Theory of Evolution). London: Cambridge University Press. (1.3)

Pearson, E.S. (1961), Some systems of frequency curves, *Stat. Tech. Res. Group Princeton, Report No. 46.* (1.2)

Pearson, E.S. (1963), Some problems arising in approximating to probability distributions, using moments, *Biometrika*, **50**, 95–111. (1.11, 8.4)

Pearson, E.S. (1965), Tables of percentage points of $\sqrt{b_1}$ and b_2 in normal samples – a rounding off, *Biometrika*, **52**, 282–5. (1.1)

Pearson, E.S. (1967), Studies in the history of probability and statistics: XVII. Some reflections on continuity in the development of mathematical statistics, 1885–1920, *Biometrika*, **54**, 341–55. (1.1)

Pearson, E.S. and Tukey, J.W. (1965), Approximate means and standard deviations based on distances between percentage points of frequency curves, *Biometrika*, **52**, 533–46. (1.9)

Pearson, K. (1894), Contributions to the mathematical theory of evolution, *Phil. Trans. Roy. Soc.*, A, **185**, 719–810. (4.1, 4.5, 4.8)

Pearson, K. (1895), Memoir on skew variation in homogeneous material, *Phil. Trans. Roy. Soc.*, A, 186, 343–414. (1.1, 1.3, 3.1, 5.1)

Pearson, K. (1901), Supplement to a memoir on skew variation, *Phil. Trans. Roy. Soc.*, A, **197**, 443–59. (1.3)

Pearson, K. (1905), On the general theory of skew correlation and non-linear regression, *Biometrika*, **4**, 172–212. (1.1)

Pearson, K. (1916), Second supplement to a memoir on skew variation, *Phil. Trans. Roy. Soc.*, A, **216**, 429–57. (1.3, 1.5)

Pearson, K. (1923), Note on skew frequency surfaces, *Biometrika*, **15**, 222–30. (3.2)

Pearson, K. (1924a), On the moments of the hypergeometrical series, *Biometrika*, **16**, 157–62. (ex. 5.13)

Pearson, K. (1924b), On a certain double hypergeometric series and its representation by continuous frequency surfaces, *Biometrika*, **16**, 172–88. (7.3)

Pearson, K. (1924c), Note on the relationship of the incomplete β-function to the sum of the first p terms of the binomial $(a + b)^n$, *Biometrika*, **16**, 202–3. (1.4)

Pearson, K. (1925), The fifteen constant bivariate frequency surfaces, *Biometrika*, **17**, 268–314. (3.4)

Pearson, K. (1936), The method of moments and the method of maximum likelihood, *Biometrika*, **28**, 34–59. (1.6)

Peizer, D.B. and Pratt, J.W. (1968), A normal approximation for binomial, F, beta, and other common related tail probabilities. *J. Amer. Statist. Ass.*, 63, 1417–56 (part I), and 1457–83 (part II, by J.W. Pratt alone). (8.5, ex. 8.4)

Perks, W. (1947), Some observations on inverse probability, including a new indifference rule, *J. Inst. Actuar.*, 73, 285–334. (2.17)

Pinkham, R.S. (1962), An approximation to the probability integral of the gamma distribution for small values of the shape parameter, *Biometrika*, **49**, 276–8. (8.5)

Plackett, R.L. (1953), The truncated Poisson distribution, *Biometrics*, **9**, 485–8. (5.11, 6.3)

Plackett, R.L. (1965), A class of bivariate distributions, *J. Amer. Statist. Ass.*, 60, 516–22. (3.8)

Pólya, G. (1931), Sur quelques points de la théorie des probabilités, *Ann. de l'Inst. H. Poincaré*, **1**, 117–62. (10.5)

224

Pratt, J.W. (1968), see Peizer and Pratt (1968).

Pretorius, S.J. (1930), Skew bivariate frequency surfaces, examined in the light of numerical illustrations, *Biometrika*, **22**, 109–223. (2.19, 3.1)

Quenouille, M.H. (1949), A relationship between the logarithmic, Poisson and negative binomial series, *Biometrics*, **5**, 162–4. (6.6)

Raff, M.S. (1956), On approximating the point binomial, *J. Amer. Statist. Ass.*, **51**, 293–303. (8.5)

Raiffa, H. and Schlaifer, R. (1961), *Applied Statistical Decision Theory*, Harvard: Colonial Press. (5.3)

Ram, S. (1954), A note on the calculation of moments of the two-dimensional hypergeometric distribution, *Ganita*, **5**, 97–101. (7.3)

Ram, S. (1955), Multidimensional hypergeometric distribution, *Sankhyā*, **15**, 391–8. (7.3, 7.4)

Ram, S. (1956), On the calculation of moments of the hypergeometric distribution, *Ganita*, **7**, 1–5. (7.3)

Ramasubban, T.A. (1958), The mean difference and the mean deviation of some discontinuous distributions, *Biometrika*, **45**, 549–56. (1.4, 5.6)

Ramasubban, T.A. (1959), The generalised mean differences of the binomial and Poisson distributions, *Biometrika*, **46**, 223–9. (5.6)

Rao, C.R. (1957), Maximum likelihood estimation for the multinomial distribution, *Sankhyā*, **18**, 139–48. (7.6)

Rao, C.R. (1958), Maximum likelihood estimation for the multinomial distribution, with a finite number of cells, *Sankhyā*, **20**, 211–8. (7.6)

Rao, C.R. (1965), On discrete distributions arising out of methods of ascertainment, *Discrete Distributions*, 320–32. (4.4)

Rhodes, E.C. (1922), On a certain skew correlation surface, *Biometrika*, **14**, 355–77. (3.3)

Rhodes, E.C. (1925), On a skew correlation surface, *Biometrika*, **17**, 314–26. (3.4)

Richardson, L.F. (1944), The distribution of wars in time, *J. Roy. Statist. Soc., Ser. A*, **107**, 242–50. (ex. 5.17)

Rider, P.R. (1953), Truncated Poisson distributions, *J. Amer. Statist. Ass.*, **48**, 826–30. (ex. 5.16)

Rider, P.R. (1961a), Estimating the parameters of mixed Poisson, binomial and Weibull distributions by the method of moments, *Bull. Int. Statist. Inst.*, **38**, pt. 2, 1–8. (4.5, 4.6)

Rider, P.R. (1961b), The method of moments applied to a mixture of two exponential distributions, *Ann. Math. Statist.*, **32**, 143–7. (4.5, 4.6, ex. 4.4)

Rider, P.R. (1962), The negative binomial distribution and the incomplete beta function, *Amer. Math. Monthly.* **69**, 302–4. (5.11, ex. 5.15)

Rider, P.R. (1965), The zeta distribution, *Discrete Distributions*, 443–4. (5.3)

Riordan, J. (1937), Moment recurrence relations for binomial, Poisson and hypergeometric frequency distributions, *Ann. Math. Statist.*, **8**, 103–11. (5.5)

Robbins, H.E. (1948), Mixtures of distributions, *Ann. Math. Statist.*, **19**, 360–9, (4.1)

Robbins, H.E. and Pitman, E.J.G. (1949), Applications of the method of mixtures to quadratic forms in normal variates, *Ann. Math. Statist.*, **20**, 552–60. (4.1)

Robertson, C.A. and Fryer, J.G. (1970), The bias and accuracy of moment estimators, *Biometrika*, **57**, 57–66. (4.8)

Robinson, E.A. (1959), *An Introduction to Infinitely Many Variates*. London: Griffin. (4.2)

Robson, D.S. and Whitlock, J.H. (1964), Estimation of a truncation point, *Biometrika*, **51**, 33–40. (1.7)

Rolph, J.E. (1968), Bayesian estimation of mixing distributions, *Ann. Math. Statist.*, 39, 692–3. (4.5)

Romanovsky, V. (1923), On the moments of a binomial $(p + q)^n$ about its mean, *Biometrika*, 15, 410–12. (5.5, 5.6)

Romanovsky, V. (1924), Generalisation of some types of the frequency curves of Professor Pearson, *Biometrika*, 16, 106–16. (1.1, 2.1)

Roy, J. and Mitra, S.K. (1957), Unbiased minimum variance estimation in a class of discrete distributions. *Sankhyā*, 18, 371–8. (6.3)

Ruben, H. (1964), An asymptotic expansion for the multivariate normal distribution and Mills' ratio, *J. Nat. Bur. Stand.*, 68B, 3–11. (8.5)

Sampford, M.R. (1953), Some inequalities on Mills' ratio, *Ann. Math. Statist.*, 24, 130–2. (8.5)

Sampford, M.R. (1955), On the truncated negative binomial distribution, *Biometrika*, 42, 58–69. (5.11)

Sandiford, P.J. (1960), A new binomial approximation for use in sampling from finite populations, *J. Amer. Statist. Ass.*, 55, 718–22. (8.5, ex. 8.1)

Sarkadi, K. (1957), Generalised hypergeometric distributions, *Magyar Tud. Akad. Math. Kutato Int. Kozl*, 2, 59–69. (5.1)

Särndal, C.E. (1964), A unified derivation of some non-parametric distributions, *J. Amer. Statist. Ass.*, 59, 1042–53. (7.3)

Särndal, C.E. (1965), Derivation of a class of frequency distributions via Bayes' theorem, *J. Roy. Statist. Soc.*, Ser. B, 27, 290–300. (7.2, 7.3)

Schilling, W. (1947), A frequency distribution represented as the sum of two Poisson distributions, *J. Amer. Statist. Ass.*, 42, 407–24. (4.5)

Seal, H.L. (1952), The maximum likelihood fitting of the discrete Pareto law, *J. Inst. Actuar.*, 78, 115–21. (5.3)

Selvin, H.C. and Stuart, A. (1966), Data dredging procedures in survey analysis, *Amer. Statistician*, 20, 20–23. (1.6)

Shenton, L.R. (1949), On the efficiency of the method of moments and Neyman's Type A distribution, *Biometrika*, 36, 450–4. (6.12)

Shenton, L.R. (1950), Maximum likelihood and the efficiency of the method of moments, *Biometrika*, 37, 111–6. (5.8)

Shenton, L.R. (1951), Efficiency of the method of moments and the Gram–Charlier Type A distribution, *Biometrika*, 38, 58–73. (2.9)

Shenton, L.R. and Bowman, K.O. (1967), Remarks on large sample estimators for some discrete distributions, *Technometrics*, 9, 587–98. (4.8, 6.10)

Shenton, L.R. and Myers, R. (1965a), Comments on estimation for the negative binomial distribution, *Discrete Distributions*, 241–62. (6.10)

Shenton, L.R. and Myers, R. (1965b), Orthogonal statistics, *Discrete Distributions*, 445–58. (6.10)

Shumway, R. and Gurland. J. (1960a), Fitting the Poisson–Binomial distribution, *Biometrics*, 16, 522–33. (6.12, 6.14)

Shumway, R. and Gurland. J. (1960b), A fitting procedure for some generalised Poisson distributions, *Skand. Aktuar.*, 43, 87–108. (6.6, 6.12)

Sibuya, M. Yoshimura, I. and Shimizu, R. (1964), The negative binomial distribution, *Ann. Inst. Statist. Math.*, 16, 409–26. (7.3)

Sichel, H.S. (1947), Fitting growth and frequency curves by the method of frequency moments, *J. Roy. Statist. Soc.*, Ser. A, 110, 337–47. (1.10)

Sichel, H.S. (1949), The method of frequency moments and its application to Type VII populations, *Biometrika*, 36, 404–25. (1.10, ex. 1.8)

Siddiqui, M.M. and Raghunandanam, K. (1967), Asymptotically robust estimators of location, *J. Amer. Statist. Ass.*, 62, 930–53. (1.9)

Simon, H.A. (1955), On a class of skew distribution functions, *Biometrika*, 42, 425–40. (5.3)

Skellam, J.G. (1948), A probability distribution derived from the binomial distribution by regarding the probability of success as variable between sets of trials, *J. Roy. Statist. Soc.*, Ser. B, 10, 257–61. (5.3)

226

Skellam, J.G. (1949), Distribution of moment statistics of samples drawn without replacement from a finite population, *J. Roy. Statist. Soc.*, *Ser. B*, **11**, 291—6. (5.3)

Slater, L.J. (1960), *Confluent Hypergeometric Functions*. London: Cambridge University Press. (5.3)

Sprott, D.A. (1958), The method of maximum likelihood applied to the Poisson binomial distribution, *Biometrika*, **14**, 97—106. (6.6, 6.12)

Sprott, D.A. (1965a), Some comments on the question of identifiability of parameters raised by Rao, *Discrete Distributions*, 333—6. (4.4)

Sprott, D.A. (1965b), A class of contagious distributions and ML estimation, *Discrete Distributions*, 337—50. (6.12, ex. 6.11)

Sprott, D.A. (1970), Some exact methods of inference applied to discrete distributions, *Random Counts*, Vol. 1, 207—32. (7.6)

Staff, P.J. (1964), The displaced Poisson distribution, *Austr. J. Statist.*, **6**, 12—20. (5.3, 6.4)

Staff, P.J. (1967), The displaced Poisson distribution — Region B, *J. Amer. Statist. Ass.*, **62**, 643—54. (5.3, 6.4)

Steck, G.P. (1968), A note on contingency-type bivariate distributions, *Biometrika*, **55**, 262—4. (ex. 3.7)

Steyn, H.S. (1951), On discrete multivariate probability functions, *Kon. Ned. Akad. Wet. Proc. A.*, **54**, 23—30. (7.3, 7.4)

Steyn, H.S. (1955), On discrete multivariate probability functions of hypergeometric type, *Kon. Ned. Akad. Wet. Proc.*, *A*, **58**, 588—95. (7.3, 7.4)

Steyn, H.S. (1957), On regression properties of discrete systems of probability functions, *Kon. Ned. Akad. Wet. Proc.*, *A*, **60**, 119—27. (3.2, 7.3, 7.4)

Steyn, H.S. (1960), On regression properties of multivariate probability functions of Pearson's type, *Kon. Ned. Akad. Wet. Proc.*, *A*, **63**, 302—11. (3.2)

Steyn, H.S. and Wiid, A.J.P. (1958), On eightfold probability functions, *Kon. Ned. Akad. Wet. Proc. A*, **61**, 129—38. (7.3, 7.4)

"Student", (1907), On the error of counting with a haemocytometer, *Biometrika*, **5**, 351—5. (6.12, 6.15)

Subrahmaniam, K. (1966), A test for "intrinsic correlation" in the theory of accident proneness, *J. Roy. Statist. Soc.*, *Ser. B*, **28**, 180—9. (7.9)

Szegö, G. (1939), *Orthogonal Polynomials*. New York: Amer. Math. Society. (App. A)

Talacko, J. (1958), A note about a family of Perks distributions, *Sankhyā*, **20**, 323—8. (2.17)

Tanner, J.C. (1953), A problem of interference between two queues, *Biometrika*, **40**, 58—69. (5.12)

Taylor, B.J.R. (1965), The analysis of polymodal frequency distributions, *J. Animal Ecol.*, **34**, 445—52. (4.8)

Teicher, H. (1954a), On the convolution of distributions, *Ann. Math. Statist.*, **25**, 775—8. (4.1)

Teicher, H. (1954b), On the multivariate Poisson distribution, *Skand. Aktuar.*, **37**, 1—9. (ex. 7.3)

Teicher, H. (1960), On the mixture of distributions, *Ann. Math. Statist.*, **31**, 55—73. (4.2, ex. 4.1)

Teicher, H. (1961), Identifiability of mixtures, *Ann. Math. Statist.*, **32**, 244—8. (4.3)

Teichroew, D. (1957), The mixture of normal distributions with different variances, *Ann. Math. Statist.*, **28**, 510—12. (4.9, 8.5)

Thiele, T.N. (1931), Theory of observations, *Ann. Math. Statist.*, **2**, 165—309. (5.13)

Thomas, M. (1949), A generalisation of Poisson's binomial limit for use in ecology, *Biometrika*, **36**, 18—25. (6.6, 6.13)

Tiku, M.L. (1964), A note on the negative moments of a truncated Poisson variate, *J. Amer. Statist. Ass.*, **59**, 1220–6. (5.11)

Tippett, L.H.C. (1934), Statistical methods in textile research, Part 2; Uses of the binomial and Poisson distributions, *Shirley Inst. Res. Memoirs*, **13**, 35–72. (8.8)

Truesdell, C. (1947), A note on the Poisson–Charlier functions, *Ann. Math. Statist.*, **18**, 450–4. (5.13)

Tsao, C.K. (1965), A moment generating function of the hypergeometric distributions, *Discrete Distributions*, 75–8. (5.6)

Tukey, J.W. (1960), A survey of sampling from contaminated distributions, 448–65, in I. Olkin *et al.* (eds), *Contributions to Probability and Statistics*. Stanford: Stanford University Press. (4.9)

Tukey, J.W. (1962), The future of data analysis, *Ann. Math. Statist.*, **33**, 1–67. (1.9)

Tweedie, M.C.K. (1947), Functions of a statistical variate with given means, with special reference to the Laplacian distributions, *Proc. Camb. Phil. Soc.*, **43**, 41–9. (6.2)

Van Uven, M.J. (1925), On treating skew correlation, *Kon. Ned. Akad. Wet. Proc.*, **23**, 793–811 and 919–35. (3.7)

Wallace, D.L. (1958), Asymptotic expansions to distributions, *Ann. Math. Statist.*, **30**, 635–54. (8.2)

Wasow, W. (1956), On the asymptotic transformation of certain distributions into the normal distribution, *Numerical Analysis*, *Vol. 6*, *Proc. Sixth Symp. in Appl. Maths.*, *Amer. Math. Soc.*, 251–9. New York: McGraw–Hill. (8.4)

Weiler, H. (1965), The use of incomplete beta functions for prior distributions in binomial sampling, *Technometrics*, **7**, 335–47. (5.3)

Wicksell, S.D. (1917), On log correlation, with an application to the distribution of ages at first marriage, *Svensk. Aktuar.*, **4**, 141–61. (3.2, 3.4)

Wicksell, S.D. (1935), Expansion of frequency functions for integer variates in series, *Proc. 8th Scand. Math. Congr.*, 306–25. (5.13)

Widder, D.V. (1946), *The Laplace Transform. 2nd Princeton Mathematical Series*. Princeton: Princeton University Press. (3.2)

Wiid, A.J.B. (1957/8), On the moments and regression equations of the four-fold negative and fourfold negative factorial binomial distributions, *Proc. Roy. Soc. Edin.*, *A*, **65**, 29–34. (7.3)

Wilkins, C.A. (1961), A problem concerned with weighting of distributions, *J. Amer. Statist. Ass.*, **56**, 281–92. (4.11)

Wilson, E.B. and Hilferty, M.M. (1931), The distribution of chi-square, *Proc. Nat. Acad. Sci.*, **17**, 684–8. (8.5, ex. 8.3)

Wise, M.E. (1950), The incomplete beta function as a contour integral and a quickly converging series for its inverse, *Biometrika*, **37**, 208–18. (8.4, 8.5)

Wise, M.E. (1954), A quickly convergent expansion for cumulative hypergeometric probabilities, direct and inverse, *Biometrika*, **41**, 317–29. (8.4, 8.6)

Wishart, J. (1949), Cumulants of multivariate multinomial distributions, *Biometrika*, **36**, 47–58. (7.3, 7.4)

Woodbury, M.A. (1949), On a probability distribution, *Ann. Math. Statist.*, **20**, 311–3. (5.4)

Yakowitz, S.J. (1969), A consistent estimator for the identification of finite mixtures, *Ann. Math. Statist.*, **40**, 1728–35. (4.5)

Yakowitz, S.J. and Spragins, J.D. (1968), On the identifiability of finite mixtures, *Ann. Math. Statist.*, **39**, 209–14. (4.5)

Young, D.M. (1967), Recurrence relations between the p.d.f's of order statistics of dependent variables, and some applications, *Biometrika*, **54**, 283–92. (7.4)

Yuan, P.T. (1933), On the logarithmic frequency distribution and the semi-logarithmic correlation surface, *Ann. Math. Statist.*, 4, 30—74.   (2.13)

Yule, G.U. (1938), On some properties of the normal distribution, univariate and bivariate, based on the sums of squares of frequencies, *Biometrika*, 30, 1—10.   (1.10)

Zoch, R.T. (1934), Invariants and covariants of certain frequency curves, *Ann. Math. Statist.*, 5, 124—35.   (1.4)

Zoch, R.T. (1935), Some interesting features of frequency curves, *Ann. Math. Statist.*, 6, 1—10.   (1.5)

INDEX

('t' following a page reference refers to the list of tables and charts in Appendix B.)